Topical Conference: Chemical Engineers in Medicine

AA002664

Held at the 2024 AIChE Annual Meeting

San Diego, California, USA
27-31 October 2024

ISBN: 979-8-3313-1687-7

Topical Conference: Chemical Engineers in Medicine

Held at the 2024 AIChE Annual Meeting

San Diego, California, USA
27-31 October 2024

AIChE
The Global Home of Chemical Engineers

Printed from e-media with permission by:

Curran Associates, Inc.
57 Morehouse Lane
Red Hook, NY 12571

Some format issues inherent in the e-media version may also appear in this print version.

Copyright© (2024) by AIChE
All rights reserved.

Printed with permission by Curran Associates, Inc. (2025)

For permission requests, please contact AIChE
at the address below.

AIChE
120 Wall Street, FL 23
New York, NY 10005-4020

Phone: (800) 242-4363
Fax: (203) 775-5177

www.aiche.org

Additional copies of this publication are available from:

Curran Associates, Inc.
57 Morehouse Lane
Red Hook, NY 12571 USA
Phone: 845-758-0400
Fax: 845-758-2633
Email: curran@proceedings.com
Web: www.proceedings.com

TABLE OF CONTENTS

ARTIFICIAL INTELLIGENCE, MACHINE LEARNING AND SYSTEMS BIOLOGY APPROACHES TO ADVANCING MEDICINE

CHEMICAL ENGINEERING PRINCIPLES ADVANCING MEDICINE

ENGINEERING CANCER

NOVEL DIAGNOSTIC AND TREATMENT APPROACHES

MEDICAL DEVICES

PANDEMIC RESPONSE, PUBLIC HEALTH, AND MRNA VACCINES

PRECISION MEDICINE AND CANCER

NANOTECHNOLOGY IN MEDICINE AND DRUG DELIVERY

101a The Language of Mass Spectra: An Application to Glycomics

Ejas A. Abtheen, Changyou Chen, Sriram Neelamegham, Rudiyanto Gunawan

In the last year, there has been a lot of excitement around generative artificial intelligence, especially with the release of large language models (LLMs), such as Generative Pre-Trained Transformers (GPT). The transformer architecture behind these LLMs uses the self-attention mechanism to learn context-dependent representation (embedding) of words, where the meaning of a word depends on its context that is defined by its position and the other words and their positions in the text (Vaswani et al, 2017). Of note, this architecture has been successfully adapted to numerous applications outside natural language processing (NLP), including bioinformatics and medicine. In this work, we developed transformer-based language models of mass spectra (MS) for glycomics, called GlycoBert and GlycoBart.

Glycosylation is the enzymatic process that attaches and modifies glycans to proteins and lipids. This post-translational modification process influences protein structure, function, and stability, and therefore has many important implications in health and disease. In medicine, glycosylation is connected to different applications, from diagnostics to therapeutics and vaccine design. For example, aberrant glycosylation patterns are key markers and drivers of cancer, making them targets for innovative diagnosis and treatments. Similarly, understanding the glycosylation of viral proteins can lead to more effective vaccines. Thus, enabling accurate glycomics profiling has broad impact in medical science.

Mass spectrometry is an indispensable tool for glycomics. A mass spectrum encodes information for the structure and composition of the precursor molecules (glycans) in its mass-over-charge (m/z) peak position and intensities. In MS data analysis, the significance of a single peak is linked to the presence and intensities of other peaks. Accounting the contextual relationship across m/z peaks is crucial for deducing the composition of the precursor molecule. The interdependency among m/z peaks in mass spectra mirrors how context shapes the meaning of a word in a sentence or paragraph. The transformer architecture using the self-attention mechanism has been proven to be extremely efficacious in handling context-dependent information for various inference tasks in NLP (Vaswani et al, 2017). For this reason, we leveraged transformers so that each m/z peak can be analyzed in the full context of the entire m/z peaks in a spectrum, enabling a nuanced and accurate inference of the precursor molecule.

The first step in applying the transformer architecture to MS data is tokenization of the mass spectra into a suitable format. We formulated a tokenization procedure that transforms a given mass spectrum into a sequence of tokens, as illustrated in Figure 1. The m/z values are converted into indices by binning which produces a sequence of m/z bin indices that mirror a sequence of words in a sentence. Besides the m/z peaks, information related to experimental settings, relative retention time, and precursor mass are included in the "MS sentence". The experimental metadata include liquid chromatography type, derivatization, ion mode, ionization type, modification, ion trap, fragmentation mode, glycan type, (relative) retention time, and precursor mass. The relative retention time and precursor mass are also binned and are represented in the sentence by their bin indices. We compiled the vocabulary for MS—the set of all possible words in MS sentences—using a recent MS dataset for glycomics called CandyCrunch (Urban et al., 2023). Each word is assigned a unique token ID, allowing encoding of MS sentences as inputs for transformers.

Among the state-of-the-art transformer-based NLP models, BERT (Bidirectional Encoder Representations from Transformers) (Devlin et al., 2018) is particularly attractive for the interpretation of MS data. The bidirectional capability of BERT allows for a nuanced contextual analysis of MS peak patterns by considering both the preceding (higher intensity) and subsequent (lower intensity) peak information. Not to mention, BERT's encoder-based architecture provides flexibility in the type of inference tasks using the pre-trained model. We developed GlycoBERT, a BERT model for MS-to-glycan inference. GlycoBERT assigns an input mass spectrum into one of the possible glycan structures from the training dataset (n = 3307). As a classifier GlycoBERT is unable to perform *de novo* glycan inference, i.e. this method is unable to predict glycan structures that are not in the training dataset. To address such an issue, we further developed GlycoBART which employed an extension of BERT model for sequence-to-sequence inference, called BART (Bidirectional and AutoRegressive Transformers) (Lewis et al., 2019). GlycoBART is formulated as a conditional sequence generation problem.

For model training, we leveraged a recent dataset of tandem MS dataset for glycomics called CandyCrunch (Urban et al., 2023). CandyCrunch dataset comprises ~505K mass spectra curated from GlycoPost and the literature. After data filtering that involved removal of spectra with unreliable glycan labels and those with missing experimental metadata, we generated train and test datasets with 410K and 72K (85:15 ratio) MS2 spectra, respectively. GlycoBERT was trained using a maximum sequence length of 512 with 12 hidden layers, each having 12 attention heads, and with a hidden latent size of 768. The training was performed on a GPU server with 4 NVIDIA Quadra RTX A5000s. The structural accuracy for the test dataset reached 95.3%. In contrast, CandyCrunch model trained on the same dataset using Temporal Convolution Neural network (TCN) achieved 87.7% accuracy (Urban et al., 2023). This finding demonstrates the promise of transformer based LLM for MS analysis.

For GlycoBART, we further developed a tokenizer for glycan structures. In this case, each glycan structure is represented by its antennae. For example, the structure "Neu5Ac(a2-3)Gal(b1-3)[Neu5Ac(a2-6)]GalNAc" is decomposed into "Neu5Ac(a2-3)Gal(b1-3)GalNAc Neu5Ac(a2-6)GalNAc". Subsequently, the monosaccharides and the linkages were tokenized into individual words using a bespoke glycan vocabulary. Here, we curated 67 tokens for monosaccharides and linkages. Our assessment showed that GlycoBart generated highly accurate predictions, reaching 93.5% test accuracy. More importantly, GlycoBart is able to perform *de novo* glycan inference for spectra and glycans that do not exist in the training dataset. This generative capability is a significant advance over methods like CandyCrunch and GlycoBert.

References:

Devlin, J., Chang, M.-W., Lee, K., & Toutanova, K. (2018). BERT: Pre-training of Deep Bidirectional Transformers for Language Understanding. *arXiv e-prints*, 1810.04805.

Lewis, M., Liu, Y., Goyal, N., Ghazvininejad, M., Mohamed, A., Levy, O., Stoyanov, V., & Zettlemoyer, L. (2019). BART: Denoising Sequence-to-Sequence Pre-training for Natural Language Generation, Translation, and Comprehension. *arXiv e-prints*, 1910.13461.

Urban, J., Jin, C., Thomsson, K. A., Karlsson, N. G., Ives, C. M., Fadda, E., & Bojar, D. (2023). Predicting glycan structure from tandem mass spectrometry via deep learning. *bioRxiv*, 544793.

Vaswani, A., Shazeer, N., Parmar, N., Uszkoreit, J., Jones, L., Gomez, A. N., Kaiser, L., & Polosukhin, I. (2017). Attention Is All You Need. *arXiv e-prints*, arXiv:1706.03762. https://ui.adsabs.harvard.edu/abs/2017arXiv170603762V

Figure Legend:

Figure 1. Language models of Mass Spectra. Mass spectra from tandem mass spectrometry are parsed into MS sentences and subsequently, these are tokenized to provide inputs for GlycoBert and GlycoBart models. GlycoBert performs a sequence classification where a MS2 spectrum input is assigned a glycan label (one out of a total of 3307 labels). In contrast, GlycoBart performs conditional sequence generation where a glycan sequence (structure) is generated given the input of MS2 spectrum.

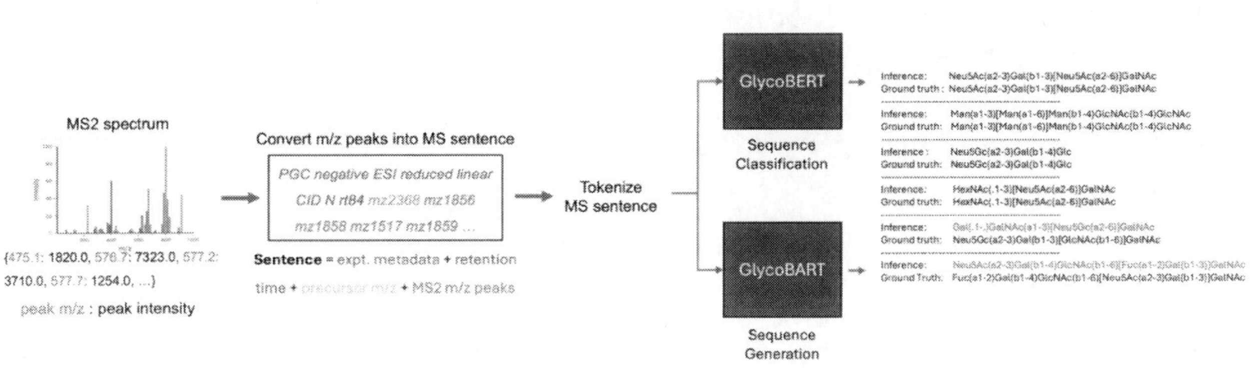
MS2 spectrum
{475.1: 1820.0, 576.7: 7323.0, 577.2: 3710.0, 577.7: 1254.0, ...}
peak m/z : peak intensity
Convert m/z peaks into MS sentence
PGC negative ESI reduced linear CID N rt84 mz2368 mz1856 mz1858 mz1517 mz1859 ...
Sentence = expt. metadata + retention time + precursor m/z + MS2 m/z peaks
Tokenize MS sentence
GlycoBERT
Sequence Classification
GlycoBART
Sequence Generation
Inference: NeuSAc(a2-3)Gal(b1-3)[NeuSAc(a2-6)]GalNAc
Ground truth : NeuSAc(a2-3)Gal(b1-3)[NeuSAc(a2-6)]GalNAc
Inference: Man(a1-3)[Man(a1-6)]Man(b1-4)GlcNAc(b1-4)GlcNAc
Ground truth: Man(a1-3)[Man(a1-6)]Man(b1-4)GlcNAc(b1-4)GlcNAc
Inference : Neu5Gc(a2-3)Gal(b1-4)Glc
Ground truth: Neu5Gc(a2-3)Gal(b1-4)Glc
Inference: HexNAc(.1-3)[NeuSAc(a2-6)]GalNAc
Ground truth: HexNAc(.1-3)[NeuSAc(a2-6)]GalNAc
Inference: Gal(.1-.)GalNAc(a1-3)[Neu5Gc(a2-6)]GalNAc
Ground truth: Neu5Gc(a2-3)Gal(b1-3)[GlcNAc(b1-6)]GalNAc
Inference: NeuSAc(a2-3)Gal(b1-4)GlcNAc(b1-6)[Fuc(a1-2)Gal(b1-3)]GalNAc
Ground Truth: Fuc(a1-2)Gal(b1-4)GlcNAc(b1-6)[NeuSAc(a2-3)Gal(b1-3)]GalNAc

101b Quantification and Machine Learning Prediction of Immunological Desensitization Biomarker

Kazeem B. Olanrewaju, Franck Kouoba

Allergic diseases pose a significant health burden globally. Immunotherapy offers a promising approach to alleviate symptoms and potentially cure allergies. However, current methods to monitor desensitization, the process by which the immune system becomes less sensitive to allergens, are subjective and lack precision. This research project aims to develop a quantitative and objective approach to assess immunological desensitization using machine learning. We will investigate the identification and quantification of immunological biomarkers that correlate with desensitization progress. Machine learning algorithms will be trained on patient data to predict desensitization outcomes and guide treatment decisions. This project has the potential to revolutionize allergy diagnosis and treatment by providing a data-driven approach to monitor desensitization and personalize immunotherapy regimens.

101c AI Driven Immune Mapping of Breast Cancer

rawlani@uW.edu Rawlani, Abishek Sankaranarayanan, Anum Kazerouni, Savannah C. Partridge, Shachi Mittal

Background: Triple negative breast cancer (TNBC) is an aggressive subtype of breast cancer. Additionally, no molecular targets are available for this cancer subtype, restricting the treatment regimen to predominantly standard chemotherapy. Neoadjuvant chemotherapy (NAC) is typically given prior to surgery to reduce tumor burden in patients with locally advanced disease. However, only a small percentage of patients show a pathological complete response after treatment. Prior research has demonstrated that the percentage of stromal (collagen) tumor-infiltrating lymphocytes (TILs), a class of immune cells, correlates to treatment response. The current clinical approaches to measure immune infiltration in tumor and associated collagen are susceptible to inter-observer variability. Additionally, the measurements are often coarse in sampling and challenging to scale up in a high throughput manner. The emergence of digital pathology, employing computational and artificial intelligence (AI) techniques, has resulted in automated and increasingly comprehensive characterization of cell populations from histopathological specimens, including quantitative measurement of stromal TILs. This enables novel analyses into the spatial interplay of immune and cancer cell populations in the tumor microenvironment. In this study, we have developed models for automated segmentation, identification, and quantification of immune, stromal, and tumor components in pre-treatment as well as surgical specimens. We have also correlated the digital immune fingerprint with the NAC response. We are also investigating whether integrating imaging and AI-based immune signatures with functional MRI features will increase the predictive power of response to treatment and recurrence-free survival.

Methods and Results: We first identified a cohort of patients with locally advanced TNBC who underwent diagnostic biopsy and breast MRI prior to NAC. We have acquired both multiplex and pathology imaging data for the immune profiling of patients. We have utilized a combination of supervised and unsupervised machine learning models to digitally segment immune cells and other diagnostic regions of the biopsy. To overcome the limitation of access to labeled data, we have utilized clustering approaches to segment the data and combine it with simpler convolutional neural networks to minimize the user-based subjective interpretation of clustering results. Additionally, we have developed a sampling strategy that minimizes the impact of patient data variability on model generalizability. We have also compared and fine-tuned deep learning-based approaches for high-throughput cell segmentation of multiplexed imaging data. Concordance correlation coefficients were calculated between automated and pathologist measurements in addition to correlating the digital signatures with patient outcome data (response to treatment and recurrence). The resulting pipeline navigates through whole slide images, segments cells, and regions of interest in an automated, high throughput, and accurate manner. Our pipeline also generates a digital stromal TILs score, as shown in the figure below, which is then used as a prognostic marker for patient stratification.

Conclusions: This work is an interdisciplinary effort combining pathology, radiology, data science, and computer vision techniques. We have developed a standardized framework for TILs assessment that can assist pathologists in consistently using this metric for patient prognosis and treatment decision-making. Finally, we have combined the power of multimodal data (three types of imaging data, patient data from medical records) with AI to develop digital indicators of patient prognosis, particularly response to treatment and recurrence free survival.

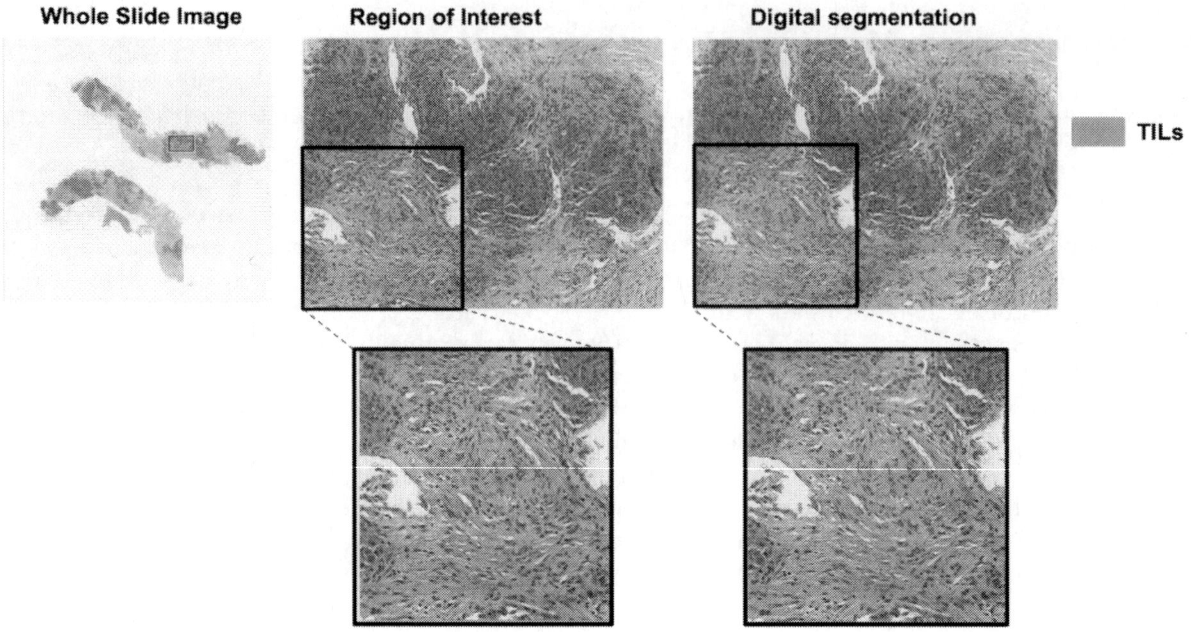

Figure 1: Digital segmentation of H&E stained whole slide image with tumor infiltrating lymphocytes marked in green.

101d Chemoenzymatic Synthesis Planning for Drugs and Bioactive Small Molecules

Hongxiang Li, Xuan Liu, Guangde Jiang, Huimin Zhao

The design of efficient synthesis routes for drug molecules plays a crucial role in pharmaceutical development. A well-designed synthesis route can significantly impact the cost, scalability, and sustainability of drug production. Moreover, it can facilitate the discovery of novel drug candidates by enabling the synthesis of complex molecules that are otherwise challenging to obtain. In this context, the development of innovative strategies for synthesis planning, such as chemoenzymatic approaches, holds great promise for advancing drug synthesis methodologies. Herein we report a reaction type score (RTscore)-guided chemoenzymatic synthesis planning (RTS-CESP) strategy for small molecules and drug synthesis. Briefly, the RTscore is trained using a convolutional neural network (CNN) to distinguish synthesis reactions from decomposition reactions and evaluate synthesis efficiency. Once multiple chemical synthesis routes are generated by a retrosynthesis tool for a target molecule, RTscore is used to rank them and find the step(s) that can be replaced by enzymatic reactions to improve synthesis efficiency. As a proof of concept, RTS-CESP was applied to 10 molecules with known chemoenzymatic synthesis routes in the literature and successfully predicted all of them, with six being the top-ranked routes. Moreover, RTS-CESP outperformed ASKCOS, a state-of-the-art chemoenzymatic synthesis planning tool, by predicting the chemoenzymatic synthesis routes for 554 out of 1000 molecules in a boutique database. Finally, RTS-CESP was employed to design a new chemoenzymatic synthesis route for the FDA-approved drug Alclofenac, which was experimentally validated and found to be shorter than the literature-reported route. Our findings indicate RTS-CESP is an effective tool for chemoenzymatic synthesis planning with strong implications for the design of efficient synthesis routes for small molecules and drugs.

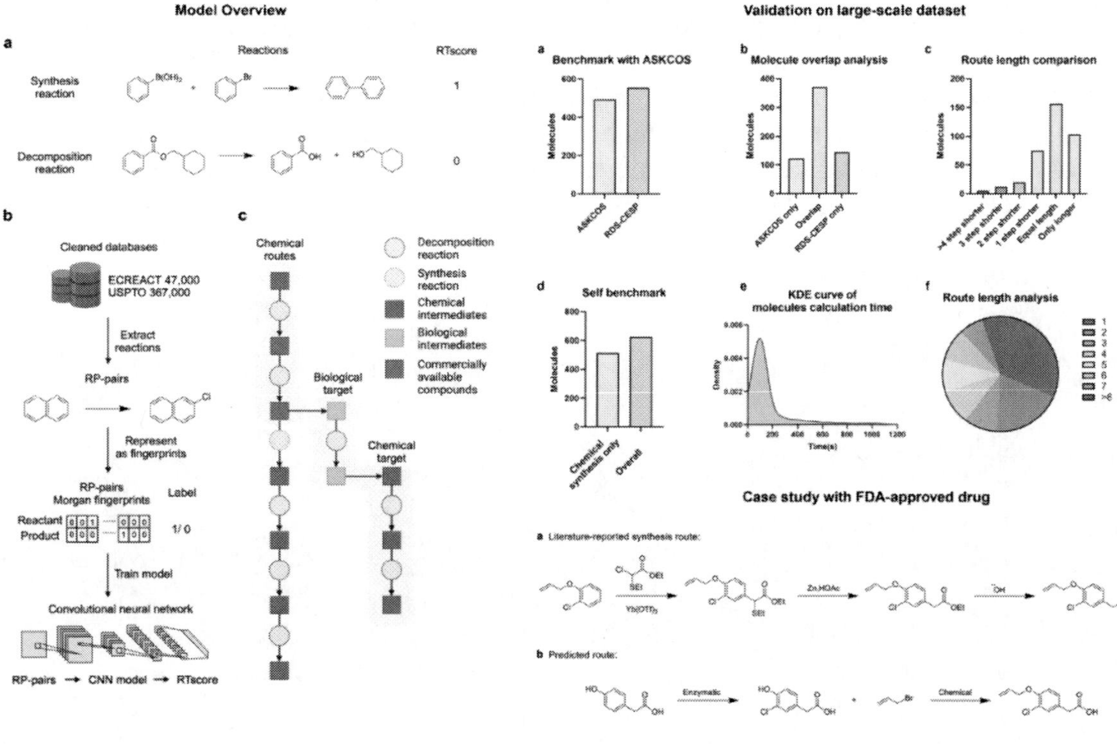

101g AI-Assisted Raman Spectroscopy for Cell Analysis

Herbert Fountain, Qiuming Yu, Shaoyi Jiang

Cellular senescence, cancers, and a wide range of other developmental and pathological processes often result in alterations in the chemical composition or metabolism of cells. The study of these changes continues to provide novel biomarkers and fundamental insights into these phenomena. The study of the dynamics and distributions of varying chemical features within cells and tissues has emerged as a powerful predictor of disease development and clinical outcomes. Despite its potential, the deployment of this analysis approach has been slow due to limits surrounding the number of metabolites and chemical features traditional methods are capable of assessing. The speed and cost of traditional methods is also a major concern. Techniques like mass spectrometry or chromatography destroy cell samples in the process of analysis, preventing further in vivo assessments which may be useful to researchers. Probe based methods are live-cell compatible but are inherently reagent dependent and are limited in terms of target specificity and multiplexing. To accelerate the pace of metabolic diagnostics we have developed a platform for the chemical analysis of single cells and bulk tissues using artificial intelligence (AI) and Raman spectroscopy. We capture intensity and wavenumber data which relay the relative concentration and chemical features found within cells that are capable of undergoing changes in polarizability (Raman active). Our platform consists of a preprocessing pipeline that removes artifacts, baseline corrects, and normalizes Raman spectra. We then apply linear (PCA) and nonlinear (UMAP) dimensionality reduction methods to sample spectra. The transformed datasets capture the variance of spectral features within individual samples while separately preserving local and global structures that exist across larger sample groups. These results are helpful in identifying unique chemical compositions within cells and may be used to discover metabolic subpopulations. Post preprocessing, we assess supervised (3 total) and unsupervised (5 total) machine learning model-candidates tasked with identifying cell types. Where applicable, we perform a grid search to find the optimal hyperparameters for each model. Platform users are then provided several performance metrics by which they can determine the optimal model for their specific study. Our best candidate model achieved 92.7% accuracy, 94% precision, and 93% recall, when discriminating between NIH3T3 and HEK293 cells. Our non-destructive, reagent-free method marks a significant step forward in metabolic diagnostics, offering detailed, scalable analysis for single cells and bulk tissues.

163a Deriving Personalized and Optimized Estrogen Dosage Profile for Optimal Endometrium Thickness in IVF Patients

Sarah Lanza, Urmila Diwekar

Introduction:

Assisted Reproductive Technologies (ART) are defined as treatments for infertility that handle eggs or embryos ("What is Assisted Reproductive Technology?"). The most commonly recommended ART treatment is in vitro fertilization (IVF), a treatment in which egg cells, also known as oocytes, are fertilized by sperm in a laboratory and then inserted back into the uterus of the patient. In vitro fertilization occurs in four stages: superovulation, egg retrieval, fertilization, and embryo transfer (Diwekar et al., 2020). In the last stage of embryo transfer or implantation of the embryo into the uterus, the embryo attaches to the endometrial lining of the uterus. A correlation exists between endometrial thickness and the success of embryo implantation. Thinner endometriums often have lower implantation success rates (Eftekhar & Mehrjardi, 2019). Daily estrogen dose is given to develop the endometrium for implantation. There are general recommendations for estrogen dosing, but the dosage is not personalized or optimized for each patient. In this work, we use the first principle-based model to link estrogen doses to endometrium thickness and a nonlinear programming (NLP) based approach to obtain a personalized and optimized dosage profile for each patient to have optimal endometrium thickness.

Objective: To personalize and optimize the estrogen dose profile to reach the optimal endometrium thickness for the endometrium preparation phase of in vitro fertilization treatment.

Model and Methods: We derive a first principle-based model and personalize it for each patient using the patient's initial days data. Then, we use an NLP optimization method to derive the optimal control profile of estrogen doses for each patient.

Expected Results: We have collected daily data from two hospitals in India on the implantation cycle for frozen embryos. The personalized model will be validated for each patient. This model will then be used to obtain the optimal estrogen profile. We will compare the thickness resulting from optimization with the suggested optimal value (8 mm at the end of the cycle). Total estrogen dosage will also be compared with the current dosage.

References:

Diwekar, Urmila, et al. "Customized modeling and optimal control of superovulation stage in in vitro fertilization (IVF) treatment." *Control Applications for Biomedical Engineering Systems*, edited by Ahmad Taher Azar, Elsevier Science, 2020.

Eftekhar, Maryam, and Sara Zare Mehrjardi. "The correlation between endometrial thickness and pregnancy outcomes in fresh ART cycles with different age groups: a retrospective study - Middle East Fertility Society Journal." *Middle East Fertility Society Journal*, 18 December 2019, https://mefj.springeropen.com/articles/10.1186/s43043-019-0013-y#citeas.

Kovacs, Gabor, and Lois Salamonsen, editors. *How to Prepare the Endometrium to Maximize Implantation Rates and IVF Success*. Cambridge University Press, 2019.

Prato, Luca Dal et al. "Endometrial preparation for frozen-thawed embryo transfer with or without pretreatment with gonadotropin-releasing hormone agonist" *Fertility and Sterility, Volume 77, Issue 5, 956 - 960*

"What is Assisted Reproductive Technology? | Reproductive Health | CDC." *Centers for Disease Control and Prevention*, https://www.cdc.gov/art/whatis.html.

163b Mechanism and Kinetic Parameters Determination of Transglycosylation of Glucose to Non-Digestible Oligosaccharide

Kazeem B. Olanrewaju

Transglycosylation, the transfer of glycosyl units between sugar molecules, offers a promising method for synthesizing non-digestible oligosaccharides (NDOs) with potential health benefits. This study investigates the mechanism and kinetic parameters of enzymatic transglycosylation of glucose to NDOs. We elucidate the specific steps involved in the reaction and identify the key enzymes responsible. Kinetic parameters, including Michaelis-Menten constant (Km) and maximum velocity (Vmax), are determined for the transglycosylation process. The influence of various reaction conditions, such as substrate concentration, enzyme loading, and temperature, is also explored. The findings provide valuable insights into the reaction mechanism and guide the optimization of NDO production through transglycosylation. This research contributes to the development of efficient and sustainable strategies for producing functional NDOs for nutraceutical applications.

163c Unraveling the Multifaceted Roles of Zinc Homeostasis in Host Defense: A Systems Biology Perspective

Sunayana Malla, Rajib Saha

The role of zinc homeostasis in pathogenesis and particularly host defense against infection, has garnered increasing attention due to its multifaceted involvement in cellular processes. This investigation helps further establish the importance of zinc's fundamental role in maintaining human health through the lens of systems biology. Zinc serves as a crucial cofactor for numerous enzymes and transcription factors involved in immune regulation, oxidative stress, epithelial barrier function, and cell viability within tissue microenvironments. Dysregulation of zinc homeostasis is associated with pathological conditions, including infection, chronic obstructive pulmonary disease (COPD), and lung cancer amongst others. Systematic approaches that integrate "omics" technologies, computational modeling, and network analyses offer a more comprehensive framework to elucidate the complex interplay between zinc dynamics, cellular function, and tissue preservation during stress. Here we hypothesized that altered zinc mobilization via ZIP8, a zinc transporter that is vital for proper immune function, will result in significant dysregulation in multiple signal transduction pathways (such as mTOR, Pi3k/Akt signaling, etc.) causing impairment of metabolic and immune responses associated with host defense. To test our hypothesis, we compared a ZIP8 knockout mouse model to wild type control mice before and after bacterial pneumonia. Using combined systems biology approaches we were able to reveal novel mechanistic insight into disease pathology as well as potential therapeutic targets to reduce the incidence of morbidity and mortality associated with pneumococcal pneumonia, a leading cause of pneumonia world-wide. The overarching goal moving forward is to develop strategies that consider dietary zinc status and related genetic susceptibilities that foster patient-specific, personalized interventions aimed at restoring zinc homeostasis in order to decrease the severity and extent of bacterial pneumonia.

163d Thermodynamic Driving Force and Protein Cost Complement the *Ccpa* Regulation Governing the Onset of *Staphylococcus Aureus* Overflow Metabolism

Nabia Shahreen, Adil Alsiyabi, Niaz Chowdhury, Vinai Chittezham Thomas, Rajib Saha

Staphylococcus aureus, a severe human pathogen, distinctively exhibits overflow metabolism at low growth rates, which is a puzzling phenotype from the bioenergetics perspective. Therefore, this study employs a synergistic experimental and modeling approach to investigate this interesting trait. First, to revisit the known suppression of the TCA cycle by *ccpA*, a *ΔccpA* mutant of *S. aureus* was created. However, the observed acetate overflow in the *ΔccpA* mutant indicated the involvement of other regulatory mechanisms. To find these regulations, we applied an in-silico analysis that identified succinate dehydrogenase, mediated by menaquinone, as a key thermodynamic bottleneck in aerobic glucose catabolism, leading to obligatory acetate overflow. Additionally, a comparative protein cost analysis of the central carbon pathway highlighted the higher resource demand of the TCA cycle over the acetate overflow pathway. Moreover, our novel algorithm, 'InteGraM' when combined with the OptMDFpathway framework, indicated that membrane crowding limits electron transport chain activity. This limitation increases dependence on acetate overflow for ATP production, a trend consistent with the membrane-real estate hypothesis. The modeling result was again verified by ^{13}C labeling experiments. Overall, this study demonstrates the importance of thermodynamic driving forces and resource allocation, in addition to *ccpA* regulation, as governing principles of *S. aureus*'s onset of overflow metabolism.

163e Probing the Spatial Proteomics of Gamma-Secretase/Notch Signaling in Breast Cancer: Coupling Correlative Microscopy with Spatial-Reaction Diffusion Modeling

Binh Nguyen, Jobin Joseph, Yangjingyi Ruan, Yueming Li, Malcolm Lane Gilchrist

The overall objective of this research is to investigate the spatiotemporal attributes of the gamma-secretase/Notch pathway and its inhibition in triple negative breast cancer (TNBC) to provide a molecular basis for therapeutic development. The mortality from TNBC in African American (AA) women is higher than in White American (WA) or European-American (EA) women. Recent studies have indicated that Notch signaling plays an important role in this racial disparity. Notch signaling has been associated with tumor transformation, proliferation, survival, angiogenesis and metastasis. We have shown that gamma-secretase cleavage of Notch is highly regulated in breast cancer cells (Villa et al (2014) 8, 1077-1092) and differs significantly in the 2D vs 3D cellular context. Our other recent studies have shown that subtle changes in membrane lipid composition affect the distribution of Notch substrates and gamma-secretase and thus enzymatic activity(Barros et al. (2020) Langmuir 36, 23, 6569–6579). At this point, ethnic differences in gamma-secretase/Notch signaling and its role in the higher incidence of poorer outcomes of AA from TNBC are poorly understood. The major question is exactly how the racial disparities would manifest. The objective of these studies is to expand the scope of our research by bringing in a systems biological modeling component to help discern the molecular underpinnings of this disparity. In essence, in the context of molecular observables of gamma-secretase/notch signaling, we aim to use modeling to search for spatial proteomics-based biomarkers to be used to guide therapeutic intervention.

We hypothesize that the characterization of the overall and intracellular spatial proteomics of g-secretase/Notch and comparison with *in silico* system biology models could yield molecular signatures of racial disparities. To interpret the underlying biophysical and kinetic underpinnings we propose to compare image-based spatial proteomics of TNBCs to spatial stochastic simulations. These models contain the gamma-secretase/Notch signaling network coded spatially onto cell structure grids obtained from the 3D confocal and EM imagery of the plasma membrane, nucleus, and other relevant intracellular structures. These models can predict the spatial distributions of the protein molecular observables in the 3D imagery, such as the build-up or depletion from cell regions under different gamma-secretase inhibitor (GSI) levels, using known GSI and enzymatic kinetic parameters. This method, in essence, produces generative models of image-based spatial proteomics that can be compared with experimental data. The major advance is that these underlying models incorporate what is known about the biophysics of gamma-secretase/Notch, in the form of membrane protein (MP) diffusivities of the enzyme and cleavage fragments that is dependent on the lipid microenvironment and also MP phase separation/localization, such as obtained in our Confocal Laser Scanning Microscopy/Atomic force microscopy (CLSM/AFM) studies of gamma-secretase/notch in supported lipid bilayers. Another important aspect of the models is that the enzyme/inhibition kinetics are integrated and the dependence of the simulated spatial proteomics on these parameters can be characterized.

This work constitutes the first-ever examination of the image-based and simulated spatial proteomics of TNBCs in the context of racial disparities in Notch and gamma-secretase pathways. We believe that this work could ultimately lead to the basis for therapeutic strategies based on gamma-secretase inhibition characteristics in the overall and spatial proteome.

163f Chemical Engineering Principles Add Clinical Value to Coronary Thermodilution Curves

Jacob Miller

Heart disease is the leading cause of death in the developed world, and occlusive coronary artery disease (CAD) has long been at the center of cardiovascular research. Although fractional flow reserve (FFR) serves as the gold standard for assessing CAD, there exists a mismatch between intervention and improved long-term patient outcomes in specific demographics. In response, researchers have begun to investigate other comorbidities that may, in tandem with FFR, better inform patient care and improve outcomes. In recent years, coronary thermodilution has been investigated for its ability to assess microvascular dysfunction, a confounding factor in the use of FFR to determine cardiovascular health and risk. However, we hypothesize that current clinical practices do not fully exploit the information available from coronary thermodilution curves and overlook potentially critical insights into coronary health. To address this problem, this research project aimed to investigate if coronary thermodilution curves could be accurately modeled with simple, mathematical models borrowing from concepts in reactor engineering and if the parameters of these models correlated with clinically relevant hemodynamic information. To this end, nine model arteries were created that spanned the design space of 20%, 50%, and 70% diameter stenoses and 30°, 45°, and 60° stenotic ramp angles. Each arterial geometry was meshed using ANSYS Mesher. The flow through each arterial geometry was simulated at 15 different flow rates ranging from 0.5 to 4.0 mL/s in ANSYS Fluent. The resulting dataset consisted of 135 unique coronary thermodilution curves, which were processed using Python. The CFD model was validated against existing literature. This study has successfully deriveda framework to recast coronary thermodilution curves to be mathematically represented as probability density functions, equal to those in both Indicator-Dilution Theory and Residence Time Distribution Theory. We then successfully applied concepts from reactor design/diagnostics in chemical engineering to derive a mathematical expression that models the thermodilution curves in terms of ideal reactor flow patterns. The model was fitted to all 135 model artery simulations with a relative absolute error across the entire cohort of 0.081, 95% CI [0.090, 0.075]. We correlated one of the fitted model parameters to the size of stagnation zones which are known to be clinically significant for their atherogenerative environment and increased risk of clotting. The Pearson correlation coefficients of the model parameter with the relative volume of stagnant blood is 0.99 for laminar flow regimes and 0.89 for turbulent flow regimes. Furthermore, we are working to correlate the model parameters with flow rate and anatomical parameters to formulate the model using only measurable quantities. Lastly, we are constructing a framework to test the model on real patient arteries and find the optimal model, still in terms of ideal reactor flow patterns, using the Sparse Identification of Nonlinear Dynamics (SINDy) algorithm. Thus far, the study has established that coronary thermodilution curves of model coronary arteries can be described with simple mathematical models. The correlation between model parameters and major flow patterns in the CFD simulations establishes that these models can be fit to coronary thermodilution curves to give a coarse-grain view of the hemodynamic environment in the artery of interest. The framework will be further developed using machine learning to better represent the primary hemodynamic patterns in real patient arteries. Our results have shown that the current implementation of coronary thermodilution curves is likely overlooking valuable information about the hemodynamic environment in the artery of interest that is known to affect patient outcomes, such as the presence of low wall shear stress (WSS) stagnant zones.

163g Protein-Loaded, Sustainably Formulated Bacterial Cellulose Nanoparticles

Gabrielle Balistreri, Eleftheria Roumeli, Elizabeth Nance

Background: Many U.S. FDA approved materials in nanomedicine technologies involve chemically intensive syntheses and are difficult to manufacturer at the kilogram scale. Our prior work used copolymer poly(lactic-*co*-glycolic acid) (PLGA)-polyethylene glycol (PEG) to deliver drugs to the injured newborn brain. Although PLGA-PEG is a viable and effective platform, the synthetic material is limited by scalability and sustainability. Green engineering is one way to address the concerns of sustainability in formulation of nano-based delivery systems. Here, we develop bacterial cellulose nanoparticles (BCNPs) for sustainable drug delivery applications and motivated by the material's bio renewable sourcing and biocompatibility, which are important properties for nanomedicine applications. BCNPs are a therapeutically relevant formulation that has a reduced environmental impact, an overall eco-friendly life cycle (Figure 1) and can be implemented following greener engineering methodologies.

Methods: We applied surface functionalizations to BC to tune the surface functionality and charge density. To achieve the modifications, we utilize literature reported strategies, such as methylation, acetylation, and amination. We first used commercially manufactured alpha cellulose and surface modified via methylation and acetylation to yield a greater hydrophobic material, while amination was used to yield a more hydrophilic material. After validating the functionalized alpha cellulose for positive and negatively charged cellulose matrices with Fourier transform infrared spectroscopy (FTIR) and X-ray crystallography (XRD), we next applied the same functionalization methods to BC pellicle. The BC pellicle was first collected from a kombucha co-culture media (bacteria and yeast) from the air and water interface, washed with sodium hydroxide and water for isolation and purifications steps, and then dissolved in small volumes of dimethylacetamide and lithium chloride. The BC pellicle was then functionalized using the same methods applied to alpha cellulose. After functional groups were confirmed, the BC pellicle was processed to create a small library of functionalized BCNPs for protein loading applications. The dissolved BC pellicle underwent a double emulsion formulation with polysorbate 80 surfactant solution to produce BCNPs. To make sure the functional groups were retained when making BCNPs from functionalized BC pellicle, we again performed FTIR and XRD. Slightly changing the pH during or post modification of the BCNPs allowed greater particle dispersion upon drying and redispersion in water. The particle size, zeta-potential, and polydispersity index (PDI) were measured by dynamic light scattering, NanoSizer, and electron microscopy.

Cytotoxicity of BCNPs were evaluated in our established organotypic whole hemisphere (OWH) brain slice platform prepared from postnatal day 10 (P10) rat brains and compared to chitosan nanoparticles and alginate nanoparticles, field standards for polysaccharide-based nanoparticles. After plating 300 µm OWH slices on semi-permeable membranes in slice culture media, slices were cultured for 4 days in vitro (DIV). At 4 DIV, BCNPs were topically applied to slices at 3 doses: 5 µg, 25 µg, 75 µg. After 24 h of BCNPs exposure, slices were stained with propidium iodine (PI), a cell death stain, and TO-PRO-3, a positive nuclei stain. Slices were fixed in methanol and PI+ cells were imaged using confocal microscopy. A separate set of slices exposed to BCNPs were collected and RNA was isolated to perform qPCR analysis to measure cytokine response. Cellular activation markers that were assessed include: Ki67, CD68, GFAP, Synapsin, CD11b, Vim. Lastly, to demonstrate the ability to load catalase into BCNPs, we incorporated 3 drug loadings % of catalase at 5%, 10%, and 20% and measured the catalase activity by enzymatic assay to identify active drug in the BCNPs.

Results: We evaluated surface functionalization strategies in BC pellicle aiming to create functionalized nanoparticles, and benchmarked our methods against literature reported reactions where alpha cellulose was used as a validation for those reactions. Compared to alpha cellulose post-surface modifications, the BC pellicle was positively charged upon amination, while methylation, acetylation, and esterification increased the negative charge of BC, as represented in FTIR and XRD data. The surface chemistry consisted of reactions with the primary and secondary alcohol groups on the BC surface, where the surface chemistry dictated interfacial properties and interactions with other species. Double emulsion formulation allowed the development of BCNPs <100 nm in diameter, slightly negative zeta-potential, and have low PDI post surface modification of BC pellicle. These physicochemical properties align with our prior work for polymer nanoparticles that show promise for delivery of proteins to the brain. The surface modified BCNPs were compared to the chitosan, a positively charged polysaccharide, and alginate as controls for colloidal stability and protein interactions for potential drug loading applications. The biocompatibility results of BCNPs in ex vivo models show BCNPs are non-toxic at lower doses and do not initiate an inflammatory response in healthy brain slices. We also demonstrate that catalase, a low cost and readily available protein, can be incorporated onto surface modified BCNPs by utilizing positively charged or aminated BCNPs that facilitate electronic interactions with the available carbonyls on catalase. Other interactions that could occur between the BCNPs and catalase could potentially be hydrogen bonding stabilizing the platform based on the crystallinity and organization of the cellulose units observed by XRD.

Conclusion:

We formulated sub-100nm BCNPs with positive and negatively charged chemical modifications and benchmarked their protein loading and cytotoxicity in the brain against field standard alginate and chitosan nanoparticles. Our results establish a tunable and sustainably derived platform in nanomedicine for the application of protein-based delivery to the brain. The small library of functionalized BC can open opportunities for other utilizations of BCNPs and an eco-friendlier way to produce therapeutics using green engineering.

References:

1. Joseph, A., Nyambura, C. W., Bondurant, D., Corry, K., Beebout, D., Wood, T. R., Pfaendtner, J., Nance, E., *Formulation and Efficacy of Catalase-Loaded Nanoparticles for the Treatment of Neonatal Hypoxic-Ischemic Encephalopathy.* Pharmaceutics, 2021. **13**(8).
2. Balistreri, G.N., et al., *Bacterial cellulose nanoparticles as a sustainable drug delivery platform for protein-based therapeutics.* RSC Applied Polymers, 2024.
3. Tardy, B.L., et al., *Deconstruction and Reassembly of Renewable Polymers and Biocolloids into Next Generation Structured Materials.* Chem Rev, 2021. **121**(22): p. 14088-14188.
4. Nguyen, N.P., et al., *Brain Tissue-Derived Extracellular Vesicle Mediated Therapy in the Neonatal Ischemic Brain.* International Journal of Molecular Sciences, 2022. **23**(2): p. 620.
5. Foster, E.J., Moon, R. J., Agarwal, U. P., Bortner, M. J., Bras, Julien, C-E., Sandra., Chan, K. J., Clift, M. J. D., Cranston, E. D., Eichhorn, S. J., Fox, D. M., Hamad, W. Y., Heux, L., Jean, Bruno., Korey, M., Nieh, W., Ong, K. J., Reid, M. S., Renneckar, S., Roberts, R., Shatkin, J. A., Simonsen, J., Stinson-Bagby, K., Wanasekara, N., Youngblood, J., *Current characterization methods for cellulose nanomaterials.* Chemical Society Reviews, 2018. **47**(8): p. 2609-2679.

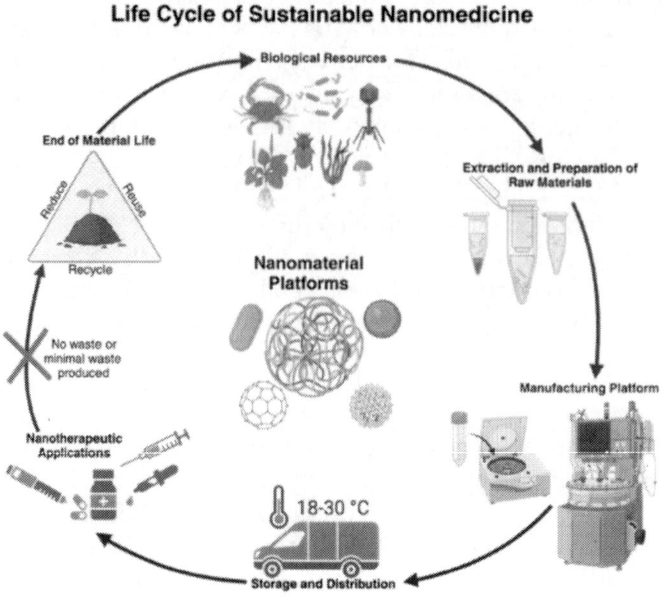

Figure 1. Environmentally friendly life cycle assessment for sustainable nanomedicine.

163h Protein-to-DNA Converter with High Signal Gain

Lin Yue Yung

DNA isothermal amplification techniques have been applied extensively for evaluating nucleic acids inputs but cannot be implemented directly on other types of biomolecules. In this work, we designed a proximity activation mechanism that converts protein inputs into DNA barcodes for the DNA exponential amplification reaction, which we termed PEAR. Several design parameters were identified and experimentally verified, which included the choice of enzymes, sequences of proximity probes and template strand via NUPACK design tool, and the implementation of hairpin lock on the proximity probe structure. Our PEAR system was surprisingly more robust against non-specific DNA amplification, which is a major challenge faced in existing formats of the DNA-based exponential amplification reaction. The as-designed PEAR exhibited good target responsiveness for three protein models with a dynamic range of 4 – 5 orders of magnitude down to femtomolar input concentration. Overall, our proposed protein-to-DNA converter module led to the development of a stable and robust configuration of DNA exponential amplification reaction to achieve high signal gain. We foresee this to enable the use of protein inputs for more complex molecular evaluation as well as ultrasensitive protein detection.

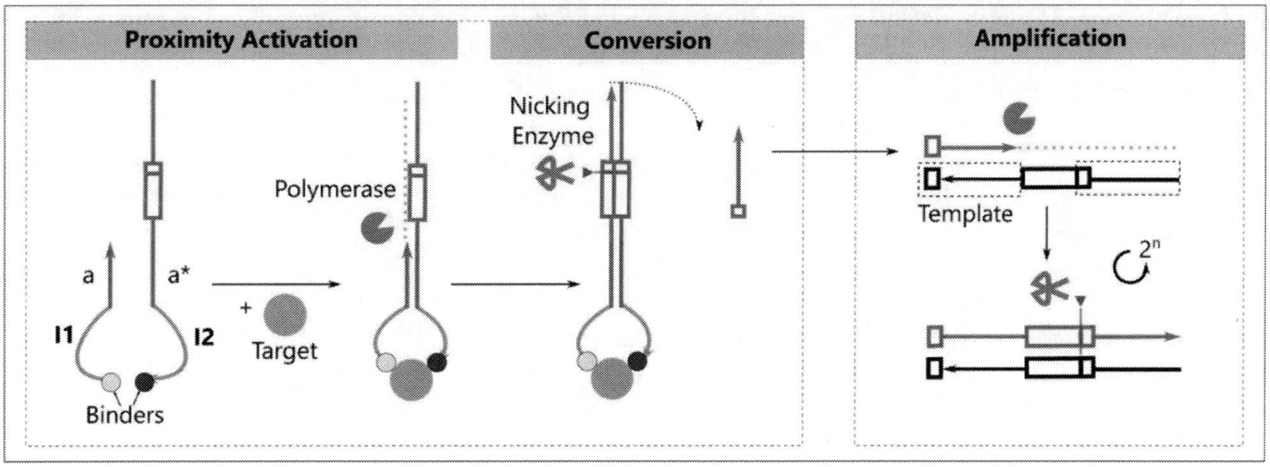

233a CRISPR-Mediated Rewiring of Epithelial-to-Mesenchymal Transition Regulatory Networks

John Nguyen, Leonidas Bleris

Cancer metastasis, defined as the proliferation of malignant cells from primary tumors to local and distant locations, remains a challenge in cancer biology and contributes to unmet clinical needs. Epithelial-to-Mesenchymal Transition (EMT) has emerged as a critical process implicated in the acquisition of invasive metastatic properties by oncogenic cells. EMT is characterized by the loss of epithelial features and the acquisition of mesenchymal traits, including enhanced motility, invasiveness, and resistance to apoptosis. Examples of key signaling pathways involved in manifestation of these various EMT phenotypes include, but are not limited to, TGFβ, Wnt/β-catenin, and Notch, and transcription factors (TFs) such as Snail, Zeb, and Twist. MicroRNAs (miRNAs), small, non-coding molecules that modulate gene expression by targeting mRNA transcripts, coordinate cellular interactions that underlie EMT phenotypes by modulating the activity of these proteins. CRISPR (Clustered Regularly Interspaced Palindromic Repeats) technology has revolutionized the field of genetic engineering. Various CRISPR-mediated DNA editing technologies, such as the type II CRISPR/spCas9 and type V CRISPR/Cas12 systems, have been extensively studied. We utilized the spCas9 endonuclease to generate stable cell models harboring distinct mutations that perturb the interactions between various miRNA to EMT-related transcription factors. These genetically stable EMT "network edge" modifications followed by thorough cell behavior characterization will contribute to our understanding of the complex network underlying the EMT and identify potential targets for therapeutic strategies that prevent the ability of cancer cells to acquire metastatic phenotypes, ultimately improving patient outcomes.

A549, human adenocarcinoma alveolar basal epithelial cells, was utilized as the EMT model. To generate mutant derivatives of the EMT model, guide RNAs (gRNAs) were designed to target and knock-out the miRNA target sites via non-homologous end joining (NHEJ) located at the 3' untranslated regions (UTRs) of EMT related transcription factors. To assess the impact of EMT network edge modification on the EMT, we perform morphological and genetic evaluation. The established EMT models were subjected to cell characterization assays to assess the impact of miRNA to EMT-TFs edge modifications. Microscopy captured morphological changes as the cell models undergo EMT. Morphological and expression profiles were observed via flow cytometry following antibody staining for key epithelial and mesenchymal surface markers. Reverse transcription-quantitative PCR (RT-qPCR) measures relative mRNA transcript level of the EMT-TFs. Immunocytochemistry (ICC) provides qualitative expression comparison of EMT related genes between the models. Proliferation and wound healing assays assess the degree of metastatic potential, their motility and growth rates, of the EMT models.

It is widely accepted that EMT related metastasis is a driver of cancer progression. The acquisition of mesenchymal cell markers is responsible for cancer metastasis, which enhance their invasive and migratory capabilities. Our results show that, through CRISPR-mediated targeted ablation of regulatory network edges between miR and key EMT-related genes, the mutant A549 cells acquire differential mesenchymal features and potentially modulating their metastatic potential. We demonstrate that targeting of miRNA to EMT-TF edges can provide useful insight on the hierarchy of specific interactions that dictate EMT and, subsequently cancer metastasis, thus advancing the field of cancer biology and a step closer to the development of cancer therapeutics.

233b JNK-Inhibition Improves Viability in Cytidine-Analog-Treated Melanoma Cells

Shayne Sensenbach, Hanny Ngo, Vahideh Angardi, Prashant Karki, Mehmet Orman

Melanoma stands as an increasingly pressing health concern within the United States. Over the past few decades, melanoma incidence has been on a steady incline. The American Cancer Society's estimates for 2024 indicate that melanoma will account for more than 8,000 deaths in the United States [1]. Notably, chemotherapy is less effective in treating melanoma compared to other cancers, and melanoma exhibits a recurrence rate of approximately 8.8% [1]. Cancer "persisters" have been recognized as a potential source of recurrent tumors [2-3]. The mechanisms behind the drug tolerance in these persister cells are complex and not well understood. Previously, we reported increased mitochondrial activity in melanoma persister cells following treatment with traditional chemotherapeutics [4]. In this study, we conducted proteomic analyses to elucidate the physiological alterations induced by chemotherapeutic agents, aiming to enhance our understanding of the underlying mechanisms of drug tolerance in melanoma.

Our initial analyses involved assessing protein expression in A375 melanoma cells (used as a model cell line) following treatment with gemcitabine (GEM) in comparison to untreated cells. Employing protein immunodetection arrays and untargeted proteomics analysis, we consistently observed an increased expression of c-Jun following GEM treatment. Moreover, the proteomics dataset identified networks of additionally upregulated and downregulated proteins (see *Figure 1*). The upregulated protein networks were related to DNA damage response (DDR), cell cycle regulation, apoptosis, the electron transport chain (ETC), mitochondrial ribosomal activity, protein folding, mRNA splicing, and multivesicular body (MVB) transport; downregulated networks included DDR, cell cycle regulation, mRNA and rRNA metabolism, tRNA aminoacylation, ribosome biogenesis, and Golgi body vesicular activity. Upon detecting the upregulation of c-Jun and its intriguing connections to the apoptotic proteins, we conducted co-treatment experiments with GEM and a JNK-inhibitor (hereafter JNK_i, which prevents c-Jun activation). Inhibiting JNK demonstrated a notable enhancement in the survival of GEM-treated cells, suggesting the involvement of c-Jun-related signaling in GEM-induced cell death. These findings were corroborated in similar experiments conducted with four additional melanoma cell lines and two other drugs falling into the same class as GEM (cytidine analogs) (see *Figure 2*). Proteomics analysis was also carried out on co-treated A375 samples (GEM + JNK_i), revealing JNK_i induced upregulations in sterol biosynthesis and cholesterol transport, as well as downregulations in cell cycle regulation and apoptosis. These findings identified multiple targets to improve chemotherapeutic treatment of melanoma, which could have broader implications for other malignancies.

References

[1] R. L. Siegel MPH, A. N. Giaquinto MSPH, and A. Jemal DVM PhD, "Cancer statistics, 2024," *CA Cancer J Clin*, vol. 74, no. 1, pp. 12–49, Jan. 2024.

[2] S. V. Sharma et al., "A Chromatin-Mediated Reversible Drug-Tolerant State in Cancer Cell Subpopulations," *Cell*, vol. 141, no. 1, pp. 69–80, Apr. 2010.

[3] M. Ramirez et al., "Diverse drug-resistance mechanisms can emerge from drug-tolerant cancer persister cells," *Nature Communications* 2016 7:1, vol. 7, no. 1, pp. 1–8, Feb. 2016.

[4] P. Karki, V. Angardi, J. C. Mier, and M. A. Orman, "A Transient Metabolic State in Melanoma Persister Cells Mediated by Chemotherapeutic Treatments," *Front Mol Biosci*, vol. 8, p. 780192, Jan. 2022.

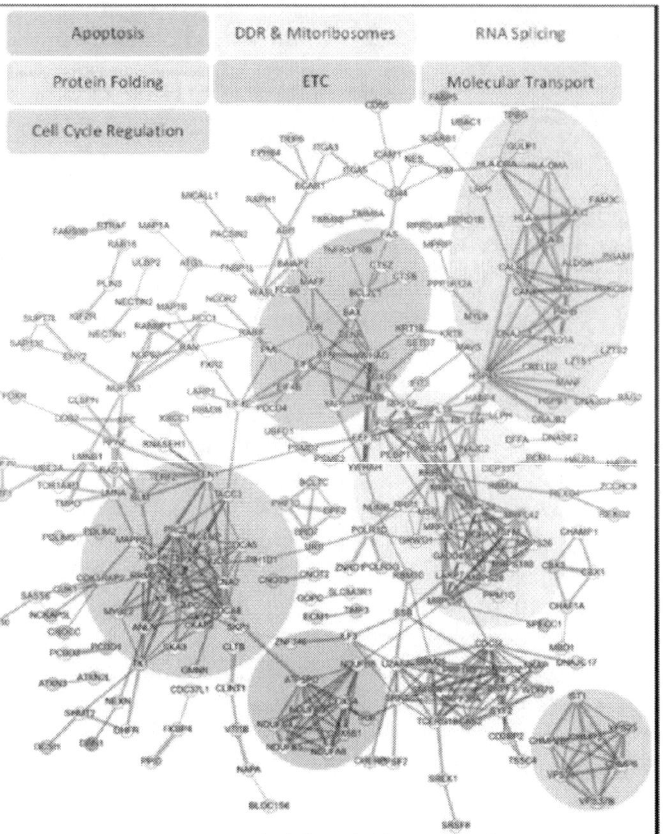

Figure 1. Gemcitabine-induced changes in protein expression in A375 melanoma cells. The highlighted protein networks were upregulated by gemcitabine treatment.

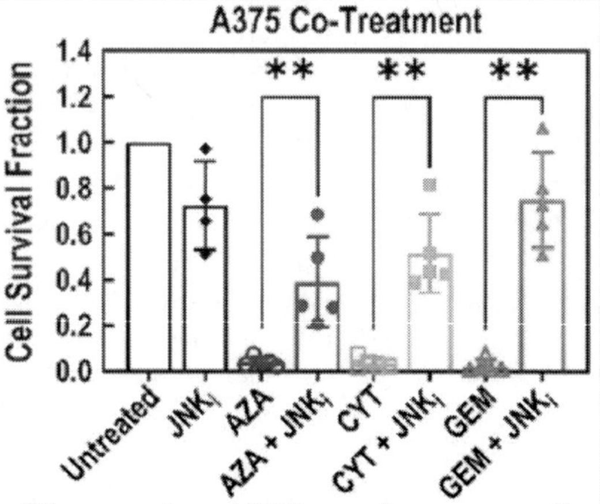

Figure 2. A375 melanoma cell viability during single and co-treatment. The viability of AZA, CYT, and GEM treated cells improves when co-treated with a JNK-inhibitor.

233c Β-Glucan Induced Innate Immune Memory Alters the Tumor Microenvironment and Enhances Checkpoint Inhibitor Therapy

Amy Laflin, Erica Wagner, Ronit Kumar, Caitlin O'Sullivan, Elyona Ihegihu, Nathan Brown, Isha Arora, Rachel Hercek, Estefani Quinones, Shaoyi Jiang

Honing the anti-tumor immune responses of patients has allowed cancer immunotherapies to be highly effective for some individuals. However, many of these therapies are unable to overcome the immunosuppressive tumor microenvironment which greatly limits their effectiveness in many solid tumors. There is a need for increasing the inflammation in the tumor microenvironment in order to increase the efficacy of FDA approved immunotherapies. Most cancer immunotherapies, including checkpoint inhibitors, help initiate and prolong the adaptive immune system's ability to kill tumor cells. There is now growing interest in developing immunotherapies that not only activate the adaptive immune system, but also harness the innate immune system's anti-tumor characteristics. The innate immune system is imperative in facilitating and maintaining a proper adaptive immune response, and innate immune cells are highly responsible for maintaining a proinflammatory or immunosuppressive tumor microenvironment. This study focused on the concept of trained immunity, also known as innate immune memory. Trained immunity is defined as the long-term functional modification of innate immune cells, allowing trained cells to have faster and stronger pro-inflammatory responses to subsequent challenges (i.e. tumor cells). The goal of this study was to determine if β-glucan peptide derived from *trametes versicolor* could increase the inflammation in the tumor microenvironment in a mouse model of melanoma. Additionally, it was determined if β-glucan peptide could augment the anti-tumor responses of the adaptive immune system and increase response rate to checkpoint inhibitors in a tumor model that is normally resistant.

β-glucan peptide was found to induce both *in vitro* and *in vivo* trained immunity. The peptide enhanced proinflammatory cytokine secretion, TNFα and IL-6, in murine macrophages when cells were restimulated with lipopolysaccharide. Additionally, bone marrow cells of mice receiving intravenous injections of β-glucan peptide showed increased levels of TNFα and IL-6 secretion upon lipopolysaccharide stimulation. A time course study was done, and seven days post-training resulted in the strongest trained immune response in the bone marrow. This experiment informed the dosing regimen in tumor models. Treatment of mice inoculated with B16F10 melanoma tumors with β-glucan peptide showed a significant reduction in tumor growth rate and increased survival compared to control mice. Flow analysis of the tumor microenvironment revealed significant reductions in the percentage of macrophages and myeloid derived suppressor cells in mice treated with β-glucan peptide. Additionally, T-cell infiltration and total numbers were analyzed using immunofluorescence staining. Staining revealed larger CD8+/Treg ratios in the tumor microenvironment of mice treated with β-glucan peptide. CD45+ immune cells in the tumor microenvironment of treated mice showed increased gene expression of proinflammatory cytokines and checkpoint molecules, indicating enhanced immune activation. Lastly, mice were co-treated with β-glucan peptide and an anti-PD1 antibody. The B16F10 melanoma model is resistant to anti-PD1 therapy, but it was found that β-glucan peptide significantly increased checkpoint inhibitor effectiveness in this model.

In conclusion, β-glucan peptide increased the pro-inflammatory responses of the innate immune cells, and it may give preference to the proinflammatory responses of myeloid cells as indicated by flow data. Not only do training agents act directly on the innate immune cells, but it appears that innate immune stimulation increases the adaptive immune system's anti-tumor responses; evidenced by immunofluorescence data and synergy with checkpoint inhibitor therapy.

233d A Multi-Scale, Optimization-Based Approach to Understanding Sulfur Fates in Integrated Biological and Transport Systems.

Carol Akpan, Lealon Martin, Emmanuel Dada

Glucoraphanin is a type of glucosinolytic phytochemical found in broccoli microgreen plants that converts to the sulfur-rich molecule sulforaphane (in the presence of the enzyme myrosinase) when the plant tissue is damaged. Studies have shown sulforaphane to be a compound with anticarcinogenic properties and other properties that generally improve metabolic health. Thus, it is proposed that higher glucoraphanin concentrations in broccoli microgreens can lead to higher sulforaphane conversion, resulting in improved anti-cancer and metabolic health maintenance from a natural, organic source.

In this work, we develop a computational approach used to identify metabolic networks (comprising known biosynthetic pathways in primary plant sulfur metabolism) that map to optimal glucoraphanin concentrations in broccoli microgreens. In our previous work, we analyzed degradation rates during the transport of harvested microgreens as a function of temperature and distance using experimental values for glucoraphanin concentration from the literature as a starting point. The current work will allow for the expansion of our previous analysis to explore degradation rates during transport as a function of variable sulfate compositions that drive primary sulfur metabolism and generate glucoraphanin concentration starting points based on identified optimal metabolic networks.

Ultimately, our proposed methodology can help us (i) to better understand underlying mechanisms for glucoraphanin production in broccoli microgreens through primary sulfur metabolism, (ii) to perform a cradle-to-grave analysis of sulfur-containing compounds by developing integrated, multi-scale, biological and transport network optimization methods, and (iii) to gain insights into new potential opportunities for optimized microgreen cultivation practices.

233e Quantitative Phase Imaging with Patient Derived Models for Cancer Treatment Optimization

Sophie Remick, Tarek E. Moustafa, Edward R. Polanco, Sandra D. Scherer, Emilio Cortes-Sanchez, Benjamin T. Spike, Bryan E. Welm, Philip S. Bernard, Thomas A. Zangle

Patient-derived organoids (PDOs) and patient-derived xenograft organoids (PDxOs) represent powerful platforms for cancer research, particularly in the development and testing of new treatments. These models provide a more accurate and clinically relevant method to study human cancer biology and therapeutic responses. PDOs and PDxOs are derived from patients' tumor cells, allowing these organoids to closely mimic the complex architecture and cellular heterogeneity of the original tumors. Because PDOs and PDxOs retain many of the biological and genetic characteristics of the original tumor, they serve as more predictive models for how a tumor might respond to specific treatments in a clinical setting. This close correlation with the patient's actual tumor response can lead to actionable predictions of clinical outcomes, providing a valuable tool for personalized medicine.

Quantitative phase imaging (QPI) is a transformative tool in oncology and functional precision medicine, particularly in the assessment of patient-derived models of cancer. QPI offers a non-invasive, label-free method to monitor real-time cellular behaviors and measure cellular dry mass in cancer cells exposed to chemotherapeutic drugs at different dosages. We will discuss the use of QPI for functional precision medicine and its advantages over traditional cytotoxicity endpoint assays such as Cell Titer Glo. We will also show the use of QPI to measure dynamic responses of cells to drugs, including assessment of heterogeneity within cellular responses.

233f Characterization of the Nano-Scale Surface Roughness of Living Cancer Cells Using a Non-Invasive Interference-Based Approach

Jose C. Contreras-Naranjo, Arul Jayaraman, Victor Ugaz

Probing the surface roughness of a living cell's membrane at the nanometer scale can offer valuable insights into its physiological condition, including any diseases present and the cell's response to treatments like anti-cancer drugs. While few methods can provide such information on live cells, atomic force microscopy (AFM) is commonly used. However, roughness characterization of cells using AFM comes with drawbacks such as limited temporal resolution and physical contact with the cell surface, which can deform delicate structures like microvilli and reveal structural aspects of the cytoskeleton rather than the membrane itself. In contrast, interference-based techniques, like reflection interference contrast microscopy (RICM), are known for their simplicity and high resolution on the nanometer/microsecond scales. RICM offers a unique non-invasive "view-from-below" perspective providing accurate information about the topography of microscopic objects near a flat and transparent surface. Here we demonstrate how RICM images of live cells can be analyzed to extract information about their nanoscale surface roughness, even when interference fringes aren't visible and fringe visibility analysis is not possible. This enables rapid and non-invasive quantification of membrane roughness at the single-cell level. Our approach relies on how a cell's rough surface scatters reflected light, as measured in an experimental RICM interferogram, relative to the scattering pattern from the corresponding smooth object, as measured from calculated intensities using appropriate cell and RICM models. We illustrate the applicability of our approach by quantifying the surface roughness of prostate cancer cells with high (PC3) and low (LNCaP) metastatic potential. Our measurements of surface roughness of living cells are consistent with current literature. Further research using blood cells and other types of cancer cells will enable the development of a liquid biopsy analysis platform, with several label-free biomarkers retrieved from RICM images, including cell roughness, employed for high-throughput non-invasive identification/classification of different cells.

233g Unmethylated Plasmid DNA Delivery Boosts Survival Against Murine Melanoma

Trishita Chowdhury, Vanshika Singh, Sudhakar Godeshala, Jordan Yaron, Kaushal Rege

Melanoma is a major healthcare burden with 1.5 million cases reported annually globally. Current therapeutic approaches include chemotherapy and immunotherapy, but low response rates and outgrowth of drug-resistant disease necessitate novel therapeutic approaches. We found that intradermal and sub-cutaneous delivery of plasmids without prokaryotic methylation ('unmethylated plasmid DNA' or 'UMpDNA') activate the innate immune system and lead to closure of full thickness wounds in healthy and diabetic, obese mice. Although transgene expression was similar with methylated plasmid DNA ('MpDNA), immune activation and wound closure were seen only with UMpDNA. Interestingly, wounding enhanced transgene expression from both MpDNA and UMpDNA in mice Building on these, we hypothesized that delivery of UMpDNA which expresses a model melanoma antigen - ovalbumin (OVA) - will activate the immune system and simultaneously act as a vaccine to combat B16F10 OVA melanoma tumors in immunocompetent mice. We hypothesized that local expression of transgene (OVA) will stimulate antigen presenting cells present in the skin, educating adaptive immune cells in regional lymph nodes, which will ultimately result in antigenic immunity. In order to test these hypotheses, delivered OVA-expressing UMpDNA or MpDNA intradermally and sub-cutaneously in C57/BL6 mice with or without 5 mm diameter splinted full-thickness wounds. DNA was administered immediately upon wound and a booster was administered on day 14 following wounding at which point, the mice were challenged with B16F10-OVA cells. Tumor growth and survival was monitored over a period of 90 days. (Figure A-D) Plasma samples (n=2) collected through submandibular bleeds showed antibodies against OVA generated after 2 weeks from the initial treatment. (Figure E) Our studies demonstrated that UMpDNA treatment resulted in significantly higher survival than seen in untreated mice (Figure F). In addition, wounding significantly enhanced survival in mice and retarded tumor progression compared to untreated mice and unwounded mice treated with OVA-UMpDNA. This study demonstrates the efficacy of plasmid DNA with unmethylated prokaryotic sites as a potential vaccine for melanoma, particularly after excision of the primary tumor.

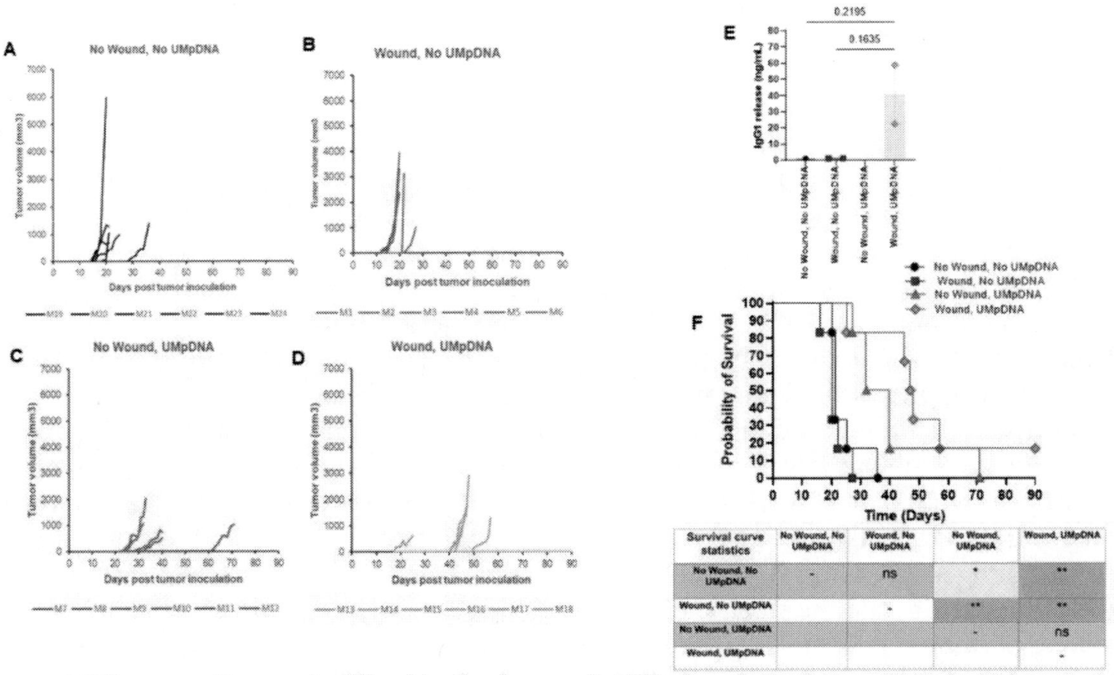

Survival curve statistics	No Wound, No UMpDNA	Wound, No UMpDNA	No Wound, UMpDNA	Wound, UMpDNA
No Wound, No UMpDNA	-	ns	*	**
Wound, No UMpDNA		-	**	**
No Wound, UMpDNA			-	ns
Wound, UMpDNA				-

A-D Tumor growth curves for different treatment groups E. IgG1 release in blood plasma F. Kaplan-Meir survival plot

291a Multichannel Acoustic Separator for High-Throughput Multiplexed Biomolecule Detection on Soft Biofunctional Particles

Cooper P. Thome, C. Wyatt Shields IV

Detection of biomarkers, including antibodies, antigens, and other physiologically relevant molecules, is essential for patient diagnosis and disease management. Multiplexed assays, which simultaneously detect multiple biomarkers in a single test, can significantly enhance diagnostic confidence. However, conventional detection assays, such as enzyme-linked immunosorbent assays (ELISA), are generally not suitable for rapid multiplexed detection, especially for biomarkers of disparate physical morphology (e.g., proteins, small molecules). Moreover, traditional biosensing techniques often require complex processing steps that introduce potential points of error and make detecting multiple biomarkers tedious. To address these challenges, we report a novel biospecific particle-based, acoustofluidics-enabled system for the simultaneous purification, preparation, and detection of multiple proteins and small molecules from complex biofluids. The central innovations of this technology are i) a class of fluorescently barcoded, functional negative acoustic contrast particles (fNACPs) that are modularly modified with antifouling polymers terminated in biorecognition motifs for the specific capture of biomarkers and ii) a 3D-printed, high-throughput multichannel acoustofluidic separator that interfaces with 12-channel pipettes for the purification of fNACPs from whole blood. Here, we show the rapid, sensitive, and simultaneous capture, purification, and detection of an antibody (Ab), antigen (Ag), and small molecule using the fNACP-based, acoustofluidic separator system.

When subjected to a half-wavelength standing wave within a microfluidic channel, positive acoustic contrast particles (PACPs) with diameters much smaller than the wavelength are forced to the pressure node, located along the middle of the channel, by primary radiation force. Conversely, NACPs are forced to the antinodes along the walls of the channel. Beyond this, particles and cells in the channel also experience a secondary radiation force, which is attractive between particles at close distances. Therefore, after acoustophoresis to the walls of the channel, NACPs subsequently aggregate. Finally, particles in an acoustofluidic channel experience a lateral acoustic radiation force that originates from gradients in the local acoustic potential along the channel and acts parallel to the channel, toward regions of local acoustic minima or maxima. At high acoustic radiation forces, this can result in acoustic trapping, which we define as the immobilization of NACPs along the walls of a straight channel in the presence of an applied flow, due to both the lateral component of the primary acoustic radiation force and the secondary acoustic radiation force exceeding the force on NACPs from Stokes' drag. In this work, we synthesized NACPs via homogenization and polymerization of a Sylgard 184 and triethoxyvinylsilane polymer precursor. Following polymerization, we isolated a low-polydispersity fraction of the NACPs by sequential vortexed filtration through 40, 30, and 20 μm filters, collecting the NACPs retained by the 20 μm filter. The surfaces of NACPs were functionalized by conjugating the particles with biotin-polyethylene glycol (PEG)-silane linkers and streptavidin, in that order. Because of its four biotin-binding sites, addition of streptavidin to the fNACPs enables subsequent conjugation of a variety of biotinylated recognition elements to allow detection of disparate biomarker types. Here, to create fNCAPs specific to a model Ab, Ag, and small molecule, we functionalized fluorescently barcoded fNACPs with biotinylated ovalbumin Ag (OVA), anti-prostate specific antigen Ab (anti-PSA), and pyrabactin resistance 1 (PYR1) protein for the detection of anti-OVA Ab, PSA Ag, and the small molecule abscisic acid (ABA), respectively. The second enabling feature of this assay is the use of a 3D-printed, high-throughput acoustofluidic separator that interfaces with 12-channel multichannel pipettes to rapidly purify captured biomolecules on the fNACPs from biofluids. The separator was designed using Fusion CAD software and printed on a

Prusa SL1S 3D printer. The device houses a 12-channel trapping array, fabricated by bonding 12 square glass capillaries to a piezoelectric transducer in parallel. The transducer is actuated by an applied AC signal at 750 kHz, which generates a pressure node along the center of each channel and antinodes along the walls of each channel. To assess trapping performance, $5x10^4$ NACPs were mixed with 1000 µL buffer, and the mixture was passed through each of the trapping channels with acoustics engaged, yielding waste and enriched samples. These samples were analyzed via flow cytometry to quantify NACP retention. To evaluate the fNACP assay performance and limit of detection (LOD), anti-OVA Ab, PSA Ag, and ABA were spiked into 100 µL buffer or porcine blood containing 10^4 fNACPs of each type (i.e., Anti-OVA, PSA, and ABA-specific) and 10^4 nonfunctionalized NACPs as an internal control. After target capture, the fNACPs and NACPs were purified using the multichannel acoustofluidic separator and incubated with fluorescently tagged secondary labels. After this incubation, the fNACPs were purified once more and analyzed via flow cytometry.

Using the described NACP synthesis and filtration approach, we isolated a low polydispersity fraction of NACPs with a mean size of ~25 µm. While the use of smaller particles in assays may offer the potential for faster particle-biomarker binding interactions due to increased diffusivity, we expected that larger particle sizes facilitate faster, more robust trapping of particles under flow, as the magnitude of acoustic radiation forces scales proportionally with the volume of particles (cubically with the radius) [2]. The discriminant forces acting on fNACPs and blood cells enabled their efficient (over 95% retention) separation in short timescales (<60 sec) using the 3D-printed, multichannel acoustofluidic separator. By functionalizing fluorescently barcoded fNACPs with a biotinylated OVA, we demonstrate the detection of anti-OVA antibodies at picomolar levels (60 pM), a sensitivity competitive with commercial ELISAs. We then show the sensitive detection of PSA by anti-PSA-functionalized fNACPs at physiological levels. Finally, we report the detection of small molecules by functionalizing fNACPs with a plant hormone receptor-inspired biorecognition motif. This third type of fNACP addresses an important limitation of traditional biosensing assays. While many immunoassays detect protein biomarkers, small molecules (i.e., molecules < 1000 Da) also serve as important indicators of disease, metabolic function, and toxin exposure. However, detecting small molecules is challenging due to their miniscule size and similarities in structure. We addressed this issue by functionalizing fNACPs with PYR1, which changes conformation when bound to the small molecule ABA, allowing complex formation with a secondary fluorescent protein, HAB1, and resultantly providing a means to quantify ABA concentration. Through barcoding the three types of fNACPs, we show simultaneous detection of all three biomarkers from whole blood using a multichannel separator that can be carried out with minimal user engagement in under 70 minutes.

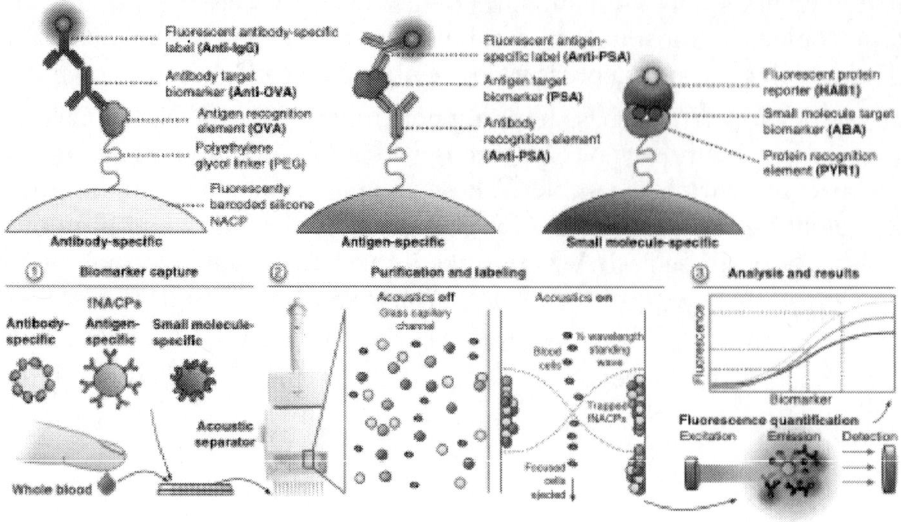
Fluorescent antibody-specific label (Anti-IgG)
Antibody target biomarker (Anti-OVA)
Antigen recognition element (OVA)
Polyethylene glycol linker (PEG)
Fluorescently barcoded silicone NACP
Antibody-specific
Fluorescent antigen-specific label (Anti-PSA)
Antigen target biomarker (PSA)
Antibody recognition element (Anti-PSA)
Antigen-specific
Fluorescent protein reporter (HAB1)
Small molecule target biomarker (ABA)
Protein recognition element (PYR1)
Small molecule-specific
① Biomarker capture
fNACPs
Antibody-specific
Antigen-specific
Small molecule-specific
Acoustic separator
Whole blood
② Purification and labeling
Acoustics off
Glass capillary channel
Acoustics on
Blood cells
½ wavelength standing wave
Trapped fNACPs
Focused cells ejected
③ Analysis and results
Fluorescence
Biomarker
Fluorescence quantification
Excitation
Emission
Detection

291b Integration of Electrochemical Sensing and Machine Learning to Detect Tuberculosis Via Methyl Nicotinate in Patient Breath

Mary Jeppson, Zachary Rasmussen, Robert Castro, Devan Jaganath, Adithya Cattamanchi, Swomitra Mohanty

Integration of Electrochemical Sensing and Machine Learning to Detect Tuberculosis via Methyl Nicotinate in Patient Breath

Background and Motivation

Tuberculosis (TB) remains a pervasive global health issue, especially in low and middle-income countries, where access to rapid and accurate diagnostic methods is often limited. The World Health Organization's End TB Strategy emphasizes the need for innovative diagnostic tools that can operate at the point-of-care and are capable of sensitivity and specificity metrics that meet or exceed the threshold for a diagnostic test. Traditional diagnostic methods, such as sputum culture and molecular testing, are resource-intensive, time-consuming, and often not feasible in resource-limited settings. In addition, the rise of drug-resistant TB strains has heightened the urgency for early and accurate detection. Recent work by our lab, the Advanced Materials and Microdevices Lab at the University of Utah, has demonstrated the potential of non-invasive, rapid, and inexpensive electrochemical sensing methods designed to specifically detect volatile organic biomarkers (VOBs) in breath for TB diagnosis. Methyl-nicotinate (MN), a VOB associated with TB, presents a promising target for such diagnostics. Our study employs a novel integration of electrochemical sensing technology and XGBoost, a machine learning algorithm, to detect MN in patient breath samples, demonstrating the potential of this method in providing an accessible, inexpensive TB diagnostic.

Methods and Results

In our clinical demonstration, conducted in Kampala, Uganda, we analyzed breath samples from a cohort of 57 individuals, comprised of 42 microbiologically confirmed TB-positive patients and 15 TB-negative controls. The detection system utilized a copper(II) liquid metal salt solution in conjunction with square wave voltammetry tailored to detect MN with high accuracy. The implementation of an XGBoost machine learning model, with optimized hyperparameters and feature extraction, allowed for the analysis of complex patterns in electrochemical data, correlating them with TB status. This approach yielded an accuracy of 78%, with a sensitivity of 71% and specificity reaching 100%. These initial results support the model's potential in distinguishing TB-positive patients from negative controls based on MN concentration in breath samples.

Significance

The preliminary results from our study represent a significant step forward in the development of accessible, non-invasive, and specific VOB detection technology for TB diagnostics. By combining electrochemical analysis with machine learning, we demonstrate the feasibility of detecting TB through breath analysis, a method that holds potential for rapid, point-of-care testing. This approach not only aligns with the WHO's goals for TB diagnosis in terms of sensitivity and specificity but also offers a scalable solution that could significantly reduce the diagnostic gap in TB-endemic regions. The success of this study paves the way for further research and development of breath-based diagnostics, with the ultimate aim of facilitating early detection and treatment initiation, thereby contributing to the global efforts to end TB.

291c Gold Leaf Electrodes for Affordable Biosensing

Marjon Zamani, Ariel Furst

The COVID-19 pandemic illustrated the urgent need for rapid and sensitive point-of-care (POC) diagnostics. Electrochemical biosensors are ideal for POC diagnostics due to their sensitivity, portability and rapid turnaround times. Gold is a common substrate for electrochemical biosensors because it can readily be modified with thiolated biomolecules. However, the high cost of gold is prohibitively expensive, making it unsuitable for accessible and affordable POC tests. To make gold electrodes more accessible, we developed gold leaf electrodes that are affordable and can be made without any specialized equipment. The thinness of the gold leaf makes our electrodes nearly an order of magnitude cheaper than their commercially available counterparts; the purity of the gold leaf allows them to outperform their commercially available counterparts, which use dopants that interfere with assay performance. We have paired our gold leaf electrodes with CRISPR-based assays to detect multiple pathogens, including human papillomavirus (HPV) from clinical samples. In doing so, we have demonstrated that these electrodes are a platform technology that can be adapted to detect nearly any infectious disease.

291d A Novel Spectroscopy-Based Diagnostic Platform for Age-Related Diseases

Herbert Fountain, Qiuming Yu, Shaoyi Jiang

The U.S. population is the oldest it has ever been, with the number of individuals aged 65 and older expected to surge from 58 million in 2022 to 82 million by 2050, marking a 47% increase. This unprecedented demographic shift is expected to coincide with a rise in patients with age-associated diseases (AADs), placing a growing burden on an already strained healthcare system. A major obstacle in treating many degenerative or age-associated diseases (AAD) is diagnosis before the onset of symptoms and irreversible progression. The foremost predictor in successfully treating any AAD or preventing related disability is an early diagnosis. However, several conditions like Alzheimer's and atherosclerosis remain difficult to detect before the advent of serious symptoms. Biological aging manifests through the systemic accumulation of molecular- and cell-level damage. The permanent growth arrest and inflammatory senescence-associated secretory phenotype (SASP) of senescent cells is a unique mechanism or hallmark by which diverse age-related pathologies arise. The accumulation of senescent cells in several non-communicable AADs suggests a potential association, highlighting the possibility of these cells to serve as biomarkers for such conditions. Accurately detecting senescent cells in culture and in complex clinical samples remains challenging due to variations in surface markers and the unreliability of traditional assays across cell types. To solve this problem, we have developed a machine learning (ML)-assisted Raman spectroscopy platform for the detection and analysis of senescent cells. We generated senescent primary human dermal fibroblasts in vitro using 3 different induction methods. We validated the development of senescence using two traditional methods, RT-qPCR and SA-β-Gal staining. Senescent cell Raman spectra were collected across different induction methods using a WITec alpha confocal Raman microscope. We chemically defined senescence in pHDFs as a ratio of normalized intensity values corresponding to major biomolecules including carbohydrates, lipids, and proteins. Our platform reliably discerns between senescent and healthy cells. Our platform consists of a preprocessing pipeline to remove artifacts, baseline correct, normalize, and reduce the dimensionality of spectral data. We then evaluated a series of supervised (ANN & CNN) and unsupervised models (LDA, QDA, SVM) as methods of classifying various cell types. We achieved accuracies greater than 90% when attempting binary classification. Our preliminary results show Raman spectra can be used to accurately identify and cluster biologically distinct cell types, potentially providing specific information about disease type and severity. We avoid the pitfalls of previous senescence detection methods which require cell-type-specific reagents or have variable degrees of reliability depending on the tissue type being analyzed. Our approach holds significant promise as both a novel tool for cell analysis and as an early AAD-diagnostic. By understanding the distinct chemical compositions of senescent cell subpopulations, we can gain valuable insight into disease progression. With the deployment of this platform, researchers will, for the first time, be able to assess the efficacy of experimental anti-aging and AAD therapeutics with chemical detail in a single-cell, non-destructive, and live-cell compatible manner.

291e A Spiral-Shaped Microfluidic Separation Channel with Integrated Microelectrodes for Cell Sorting and Position Analysis.

Yunhao Peng, Bruce K Gale, Himanshu Sant

Background and Motivation

Cell/particle soring within a heterogeneous population is a crucial preparatory step in many science and medicine fields, including hematology, immunology, health/disease diagnosis, drug development, and cancer research. Various passive cell separation microfluidic devices have been developed in the domain of microfluidics. In these devices, heterogeneous cells develop distinctive migration patterns according to their physical properties (e.g., size and shape) inside specially designed microfluidic channels while under pressure-driven flow. As a result, cells elute from the channel at different rates and positions (like liquid chromatography but with cells). These devices have many unique advantages over conventional cell separation methods (e.g., size-based filtration), including low cost, small footprint, ease of instrumentation, label-free, reduced sample waste, and high sample throughput. However, many devices still require external methods, such as high-speed camera imaging or fluorescent flow cytometry, to analyze separated cells.

In contrast, impedance flow cytometry (IFC) analyzes cells based on the impedance change caused by cells traversing across electrodes. The ease of integration into the microfluidic channel and other advantages make the IFC an ideal end-column-detection method for these devices. In this work, we demonstrated for the first time a cell sorting device combining a spiral-shaped inertial-based microfluidic channel for cell separation and integrated microelectrodes for cell counting and focusing efficiency identification. A unique electrode design was proposed to detect both cells and their lateral positions inside the channel.

Methods and Results

Polystyrene microspheres were injected into the channel, while a high-speed camera was used to record their trajectories across the electrodes. The detector output amplitude signals were extrapolated to measure the positions of individual particles. The ability of the spiral-shaped microfluidic channel to focus particles was examined against various sample flowrates by measuring individual microspheres' lateral positions. The focusing of 10 μm microspheres was around 20 μL/min flowrate. A low (500 kHz) and a high (4 MHz) frequencies were applied to the electrodes to detect Giardia and cryptosporidium cysts electrically. The difference in the electrical amplitude and phase ratio was found to be most prominent at low frequencies, demonstrating the differentiation of the two parasite types using the electrodes.

Significance

The preliminary results from our work represent an advance in microfluidic cell sorting with onboard cell detection while indicating an alternative to the gold standard flow cytometry. The proposed electrodes can be a universal detector for various microfluidic chromatography or separation systems. While having a comparably small footprint, the device is readily field deployable. For instance, it could be utilized to monitor the presence of bacteria and parasites in drinking or recreational water.

291f Anodic Stripping Voltammetry Detection of Conjugated Silver Nanoparticles As Electrochemical Probes

Shruti Hegde, Dhruv Patel, Himanshu Sant

Background and Motivation

Microbial pathogens in water pose a significant health risk. There is a need for rapid, cost-effective detection of toxins in water. Polymerase chain reaction (PCR) and enzyme immunoassay techniques are the gold standard for pathogen detection. However, these methods require several complicated processing steps and expensive instrumentation, which makes them unsuitable for use in point-of-use systems. By comparison, electrochemical biosensors are a practical solution because they can quickly, inexpensively, and sensitively detect pathogens. Silver nanoparticles (AgNPs) have been used for electrochemically detecting DNA, protein, and viruses as an alternative for enzyme immunoassay and in situ PCR. Anodic stripping voltammetry (ASV) is an electroanalytical method that uses a potential sweeping technique by stripping the electrode of the deposited metal ions present in solution at concentrations below parts per billion (ppb). Analytical responses from ASV are obtained in seconds, and the technique can be deployed in the field by eliminating the need for bulky instrumentation and elaborate sample preparation. This study will focus on a novel electrochemical biosensing method that utilizes AgNPs as electrochemical tags and readily available platinum (Pt) electrodes for detecting antibody-conjugated AgNPs for pathogen detection.

Methods and Results

A three-electrode system consisting of a platinum (Pt) working electrode, a Pt counter electrode, and an Ag/AgCl reference electrode was used for detection. Dissolved AgNPs were first deposited on the working electrode and stripped by a linear sweep voltammetry scan. Similarly, AgNPs coated with human IgG antibodies (HIgG) were stripped to determine the stripping peak of AgNPs-conjugated antibodies. Based on the experimental results, the stripping peak potential of HIgG-coated AgNPs is between 0.19 V and 0.22 V, whereas the stripping potential of AgNPs is ~ 0.14 V. Both these peaks correspond to the stripping peak observed for Ag(I) metal (0.12-0.2 V). This suggests that AgNPs' electrochemical peak would appear at a voltage similar to Ag metal and supports the use of AgNPs as electrochemical probes for use in a biosensor. The experiments were then reproduced for different antibody-AgNPs solution concentrations to validate the response's linearity over the AgNPs concentration range (0.1-10 ng/ml).

Significance

To our knowledge, no directly comparable electrochemical detection of AgNPs as probes using unmodified Pt electrodes for ASV was found in the literature. The results from this study will lay the groundwork for detecting toxin antibodies in water. The results obtained in this study have bridged the gap between ASV's use as an alternative analytical method for detecting biological components and its application as a low-cost sensing platform that can be applied in areas that otherwise require tedious sample preparation and detection approaches. We demonstrate the novel use of ASV for biosensing and establish an analytical approach that uses electrode materials that are commercially available and easily scalable for multiplexed analysis.

178aa Injectable in situ crosslinkable depots for ultra-long-acting delivery of hydrophilic therapeutics

Sohyung Lee, Jeffrey M. Karp

Patient adherence to medications is a major obstacle towards achieving effective treatment of numerous diseases, especially for chronic conditions that require treatment throughout life. Sustained release technologies, including long-acting injectables (LAIs) and implants simplify dosing schedules and enhance treatment regimen adherence, which is particularly advantageous in low resource settings with limited healthcare infrastructure.

Multiple LAIs and implants are in clinical use for the treatment and prevention of different diseases.[1, 2] Conventional LAI approaches such as microparticles,[3] *in situ*-forming implants (ISFI), and wet milled particles have been documented to achieve prolonged release of hydrophobic drugs for several months but have poor ability to achieve similar ultra-long-term release of hydrophilic drugs[4-7]. In the case of ISFI, which typically consists of a hydrophobic polymer, poly(lactic-co-glycolic acid) (PLGA) dissolved in N-methyl-2-pyrrolidone (NMP), solvent efflux during phase conversion tends to release a significant amount of drug as initial burst, which increases with the hydrophilicity of the drug.[6] Additionally, previously developed LAI approches promote significant influx and efflux of water due to their large pores, leading to fast diffusion of hydrophilic drugs.[3, 6, 8] Wet milling – a commonly employed approach for formulating LAIs[9, 10] require the drug to be hydropbohic and are fundamentally incompatible with hydrophilic drugs.[1, 11, 12] Although implantable devices have shown success for long-term delivery of both hydrophobic[13] and hydrophilic drugs[14, 15], they require invasive, time-consuming medical procedures for insertion, which may pose significant challenges, particularly in low resource settings and low-middle income countries. Furthermore, implants in general tend to be more susceptible to local inflammation when compared to injectable alternatives.[16, 17]

Hydrophilic drugs constitute a major fraction of all the drugs used for the prevention, treatment and management of chronic conditions. Examples include anti-psychotics, anti-depressants, anti-convulsants, antibiotics, and drugs for the treatment of substance abuse disorder (SUD). Although a few hydrophilic drugs have clinically approved LAI formulations, their release typically last for only a month. For example, Vivitrol®, an LAI suspension of naltrexone-loaded PLGA microspheres for treating SUD, provides 30 days of drug release,[18] which is sub-optimal since SUD often requires therapy for several years. Therefore, there is an unmet need to develop an injectable platform that enables ultra-long-term delivery of hydrophilic drugs for several months. Additionally, the platform should be designed to be biodegradable for safe breakdown and clearance from the body while also being retrievable in the event of local or systemic drug toxicity.

We report an injectable platform that addresses the aforementioned limitations of current LAI approaches, thereby resulting in ultra-long-term release of encapsulated hydrophilic drugs for at-least 6 months. This platform is derived from an ultra-low-molecular-weight liquid pre-polymer – polycaprolactone (PCL), which has been used previously in multiple FDA approved products. PCL is chemically modified with methacrylate groups, enabling it to undergo cross-linking upon injection. The liquid pre-polymer can effectively suspend or dissolve both hydrophilic and hydrophobic drugs, and can be easily injected through a standard 18-23 gauge needle. Upon co-injection with clinically used radical initiator and accelerator, benzoyl peroxide (BPO) and N,N-dimethylparatoluidine

(DMT), the pre-polymer mixture undergoes a time-dependent (1-10 minutes) cross-linking *in situ*. The cross-linking process involves radical polymerization of methacrylated PCL creating new carbon-carbon bonds. This process encapsulates the drugs physically, creating a solid monolithic structure hereafter referered to as an *in situ* cross-linked depot (ISCD). Hydrolysis of the polymer ester bonds allows gradual erosion of the depot over time, obviating the need for surgical removal after depot exhaustion, and ensuring safe clearance from the body.

ISCD has two key features, which enable it to achieve ultra-long-term release of hydrophilic drugs. These include a solvent-free design and a dense mesh network, both attributed to the use of ultra-low-molecular weight PCL. The liquid state of the pre-polymer obviates the need for a solvent, minimizing the risk of high burst release, commonly associated with solvent exchange processes in ISFI.[7] Being ultra-low molecular weight, methacrylated PCL forms a dense mesh upon cross-linking, which limits water influx/efflux and hence the drug release.

We have demonstrated sustained release of multiple hydrophilic drugs with varying water solubilities ranging from 5.63-112 mg/mL and belonging to a diverse class of therapeutics, including anti-retrovirals, opioid antagonists and antibiotics. ISCD showed sustained release of all the drugs for at least 6-10 months *in vitro*, suggesting the versatality of the platform. We also demonstrated ultra-long-term release of two proof-of-concept hydrophilic drugs - tenofovir alafenamide (TAF) and naltrexone (NAL) *in vivo* in rats. A single subcutaneous injection of ISCD formulations loaded with TAF or NAL resulted in sustained plasma concentration for at least 6-7 months. ISCD outperformed the conventional ISFI platform in TAF delivery, exhibiting notably lower burst release compared to ISFI, and providing a prolonged drug release duration of up to 7 months. ISCD also showed ultra-long-term release of a hydrophobic drug – Tacrolimus (TAC) for at least 6 months in rats. Pharmacokinetic (PK) modeling predicted sustained delivery of NAL and TAC in humans, achievable with ISCD. We also identified design parameters that can tailor the polymer network to tune drug release kinetics of ISCD. Notably, modulating the intrinsic factors, including decreasing the concentrations of BPO and DMT, or using a higher molecular weight of methacrylated PCL results in a lower cross-linking density within the polymer network, thereby increasing the drug release. Additionally, integrating external polymeric additives such as polyethylene glycol (PEG) alongside methylacrylated PCL can modulate the depot's hydrophilicity, increasing drug release. External additives with varying degree of methacrylation (mono, di or tri) can also modulate drug release by altering the cross-linking density of the polymer network. Finally, we also demonstrated the feasibility of achieving ultra-long-term release of clinically relevant combination regimens of hydrophilic drugs, encapsulated in a single depot, and demomstrated the biocompatibility and retrievability of ISCD.

Together, our findings underscore the potential of ISCD as a versatile platform to enable ultra-long-term delivery of both hydrophilic and hydrophobic drugs. This approach has holds potential to revolutionize therapy options across a variety of chronic conditions where patient adherence is critical. To our knowledge, this is the first report demonstrating ultra-long-term delivery of hydrophilic drugs using an injectable system.

1. Li, W. et al. Clinical translation of long-acting drug delivery formulations. *Nature Reviews Materials* **7**, 406-420 (2022).
2. Nkanga, C.I. et al. Clinically established biodegradable long acting injectables: an industry perspective. *Advanced drug delivery reviews* **167**, 19-46 (2020).
3. Town, A.R. et al. Tuning HIV drug release from a nanogel-based in situ forming implant by changing nanogel size. *Journal of materials chemistry B* **7**, 373-383 (2019).
4. Goonoo, N. et al. Naltrexone: A review of existing sustained drug delivery systems and

emerging nano-based systems. *Journal of Controlled Release* **183**, 154-166 (2014).
5. Benhabbour, S.R. et al. Ultra-long-acting tunable biodegradable and removable controlled release implants for drug delivery. *Nature communications* **10**, 4324 (2019).
6. Joiner, J.B. et al. Effects of drug physicochemical properties on in-situ forming implant polymer degradation and drug release kinetics. *Pharmaceutics* **14**, 1188 (2022).
7. Kempe, S. & Mäder, K. In situ forming implants—an attractive formulation principle for parenteral depot formulations. *Journal of controlled release* **161**, 668-679 (2012).
8. Maturavongsadit, P. et al. Biodegradable polymeric solid implants for ultra-long-acting delivery of single or multiple antiretroviral drugs. *International Journal of Pharmaceutics* **605**, 120844 (2021).
9. Bhakay, A., Merwade, M., Bilgili, E. & Dave, R.N. Novel aspects of wet milling for the production of microsuspensions and nanosuspensions of poorly water-soluble drugs. *Drug development and industrial pharmacy* **37**, 963-976 (2011).
10. Loh, Z.H., Samanta, A.K. & Heng, P.W.S. Overview of milling techniques for improving the solubility of poorly water-soluble drugs. *Asian journal of pharmaceutical sciences* **10**, 255-274 (2015).
11. Chaudhary, K., Patel, M.M. & Mehta, P.J. Long-acting injectables: current perspectives and future promise. *Critical Reviews™ in Therapeutic Drug Carrier Systems* **36** (2019).
12. Wright, J.C. & Burgess, D.J. Long acting injections and implants. (Springer, 2012).
13. Sitruk-Ware, R. New progestagens for contraceptive use. *Human reproduction update* **12**, 169-178 (2006).
14. Li, L. et al. Long-acting biodegradable implant for sustained delivery of antiretrovirals (ARVs) and hormones. *Journal of Controlled Release* **340**, 188-199 (2021).
15. Gunawardana, M. et al. Fundamental aspects of long-acting tenofovir alafenamide delivery from subdermal implants for HIV prophylaxis. *Scientific reports* **12**, 8224 (2022).
16. Kastellorizios, M., Tipnis, N. & Burgess, D.J. Foreign body reaction to subcutaneous implants. *Immune Responses to Biosurfaces: Mechanisms and Therapeutic Interventions*, 93-108 (2015).
17. Su, J.T. et al. A subcutaneous implant of tenofovir alafenamide fumarate causes local inflammation and tissue necrosis in rabbits and macaques. *Antimicrobial agents and chemotherapy* **64**, 10.1128/aac. 01893-01819 (2020).
18. Bartus, R.T. et al. Vivitrex®, an injectable, extended-release formulation of naltrexone, provides pharmacokinetic and pharmacodynamic evidence of efficacy for 1 month in rats. *Neuropsychopharmacology* **28**, 1973-1982 (2003).

355a 3D Spheroid Stability and Morphology on Elastin-Based Coatings

Sheetal Chowdhury, Seth Lenoir, Solange Tchounwou, Joshua Speed, Gene Bidwell, Amol Janorkar

Objective: Current *in vitro* adipose cell culture models do not replicate progressive adiposity due to limitations in facilitating the gradual intracellular fat buildup in 3D clusters of adipose-derived stem cells (ASCs) over extended periods. Previously, we cultured human ASCs as 3D spheroids on elastin-like polypeptide-polyethyleneimine (ELP-PEI) coatings and ELP-PEI coatings containing Arginine-Glycine-Aspartic Acid (RGD) residues. We observed that adding three RGD residues to the C-terminus of ELP showed superior spheroid retention. In this study, we qualitatively and quantitatively assessed the impact of the RGD-containing ELP-PEI coating on adipocyte morphology and differentiation over an eight-week culture period.

Materials and Methods: ELP conjugated with RGD was produced from genetically modified *Escherichia coli* bacteria via suspension culture. These ELP-RGD molecules were conjugated with PEI (MW= 800 Da) via carbodiimide chemistry. 5 mg/mL of 5 mol% ELP-$(RGD)_3$-PEI conjugates were adsorbed onto 24-well tissue culture polystyrene (TCPS) plates by placing 200 μL solution in each well. 50,000 human ASCs per well were cultured for 3 days on coated plates to form 3D spheroids. The spheroids were differentiated along the adipogenic lineage for 3 days and matured for 8 weeks.

Results: We observed several dynamic changes in cell morphology over the cell culture period. Initially, on Day 3, the cells aggregated to form spheroids and maintained the spheroid morphology after 3 days of differentiation. After 7 days of maturation in adipogenic media, it was observed that the spheroids were surrounded by a monolayer of 2D cells. By day 14, there was a shift toward a predominantly 2D morphology, although with scattered cell clusters. Interestingly, by day 28, there was a reformation of spheroids, indicating a cyclic pattern of spheroid formation and dispersal within the culture period. The spheroids then continued to increase in size till day 56.

Conclusion: This observation underscored the dynamic nature of cell-surface interactions of the human stem cells with the elastin-based coatings containing RGD. We will further probe these results by measuring the adipocyte gene expression and integrin binding.

355b All-Polymer-Based Configurable Intracortical Neural Electrode Array for Chronic In Vivo Brain Neural Recording

Seoyeon Won, Huanan Zhang

Implantable neural electrode arrays are promising bioelectronic platforms for recording electrophysiological signals from brains with high precision. However, most commercial electrode arrays have a large mechanical mismatch with the brain tissue and offer limited configuration due to material choices and conventional semiconductor fabrication processes. The development of next-generation tissue-compliant electrode arrays requires the convergence of advanced fabrication processes and novel bioelectrically active materials. In this study, we introduce the concept of integrating advanced two-photon 3D-printed technology and tailored conductive polymer electrodeposition to create the first functional all-polymer electrode arrays. The two-photon 3D printing technology is used to create 3D-printed microchannels to configure neural electrode arrays on demand. Then, an optimized conductive polymer deposition protocol (poly(3,4-ethylenedioxythiophene) doped with tetrafluoroborate (PEDOT:BF_4)) is used to create 3D bioelectrically active materials inside 3D-printed microchannels. These self-assembled PEDOT:BF_4 all polymer neural electrodes can reduce the mechanical mismatch between electrodes and brain tissue.

Moreover, they are electrochemically stable and have low interfacial impedance under physiological conditions. The results of impedance measurements reveal a lower impedance of the PEDOT electrodes than conventional metal and semiconductor electrodes. A chronic animal study (6 weeks) was also implemented to conduct electrophysiology recording experiments in free-moving animals. The chronic electrophysiology experiments have demonstrated stable recording of single-unit neural activities over 6 weeks. This is the first demonstration of chronically functional all-polymer intracortical neural electrode arrays for single-unit neural recording. These results indicate that the 3D-printed neural arrays based on templated PEDOT:BF_4 are a promising next-generation bioelectronic platform for intracortical recording of electrophysiological signals. Moreover, the numerous designs available through 3D printing offer opportunities to go beyond the current neural array technology.

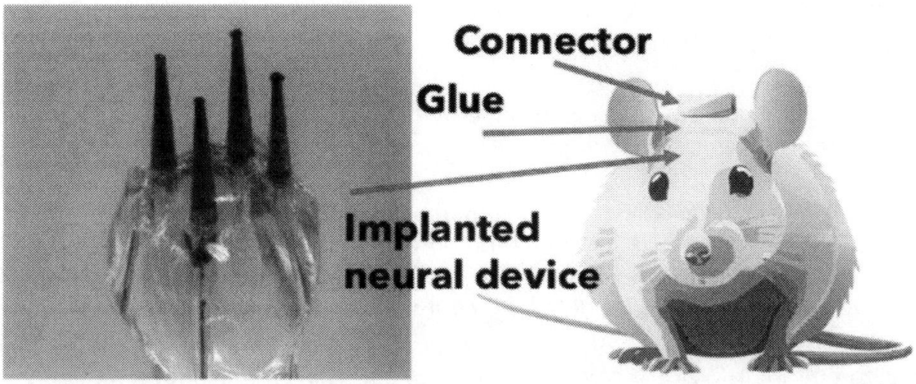
Connector
Glue
Implanted
neural device

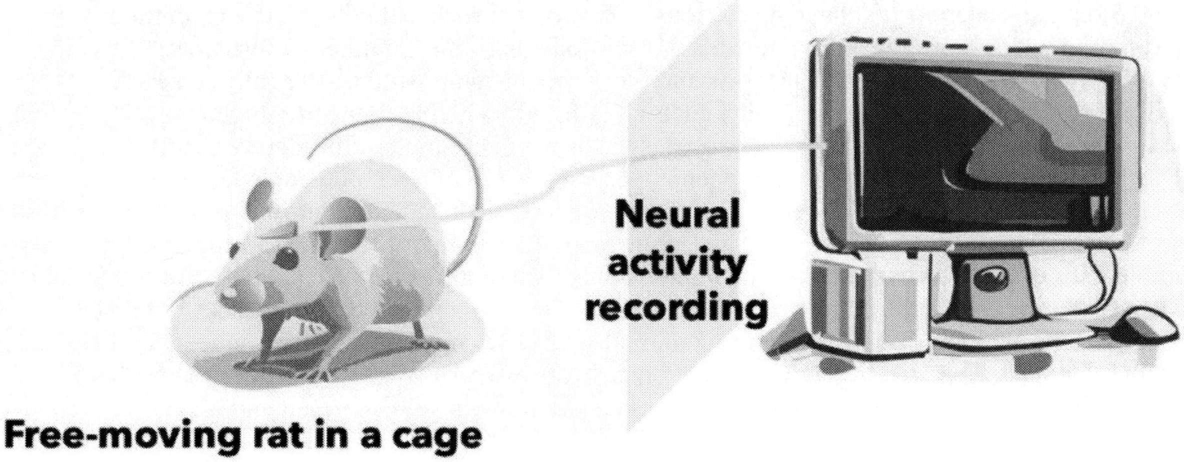
Neural
activity
recording
Free-moving rat in a cage

355c 3D Mibrain-on-Chip for Modeling BBB Function and Delivery to the Brain Inclusive of All Major CNS Cell Types

Alice Stanton, Alan Jiang, Rebecca L. Pinals, Li-Huei Tsai, Robert Langer

Treatment development for neurological disease has been limited by the tight and highly selective human blood brain barrier (BBB). A major bottleneck to developing effective therapeutics is that current models fail to recapitulate human brain tissue and its BBB with sufficient fidelity. While animal models have been a helpful research tool, the rodent BBB is far less selective than human and does not mimic the complexity of human genetics. *In vitro* models have the advantage of enabling high throughput screening, real-time monitoring at high spatiotemporal resolution, and readily tunable genetic backgrounds in combination with iPSC technology. The brain is comprised of six major different cell types arranged in a stereotyped 3D architecture, each of which affects BBB properties. Microphysiological systems harnessing natural biomaterials have enabled the co-culture of 3 of the relevant cell types into BBB models for a reductionist model. However, these models are missing the contributions of neurons and other cells that strengthen the BBB, have not yet demonstrated selective barrier properties matching human brain tissue on the molecular level, and often use materials that have deleterious effects on neurons and inhibit their co-culture. Engineered biomaterials provide highly tunable 3D scaffolds that can promote the formation of tissue mimics with physiologically relevant stiffness and biochemical compositions. Recently, we have engineered a soft hydrogel biomaterial, Neuromatrix Hydrogel, with brain-specific biochemical cues that supports the self-assembly of all six brain cell types to form a 3D immuno-glial-neurovascular miBrain model with enhanced cell- and tissue-scale phenotypes (**Fig. 1A-F**). Given that engineered hydrogels have limitations for 3D cell culture in confined microfluidic environments due to their innate swelling properties that introduces deleterious stresses to cells, we engineered a novel microfluidic platform, GelChip, to be compatible with engineered hydrogels and complex 3D cell co-cultures harnessing a 3D Printing strategy (**Fig. 1G**). GelChips enable the culture of miBrains with 3D perfusable vascular networks within co-cultures of all six major brain cell types arranged in integral BBB and neurovascular units (**Fig. 1H**), microglial immune cells (**Fig. 1I**), and myelinated neuronal networks (**Fig. 1J**), to form the miBrain-on-Chip. This is the first perfusable brain model to our knowledge with iPSC-derived endothelial cells in combination with other iPSC-derived cell types, here with six iPSC-derived cell types. With the miBrain-on-Chip, we have characterized the barrier properties, assessed delivery into the parenchymal brain tissue through the BBB, and harnessed the model to probe the effect of *APOE4*, the strongest known genetic risk factor for sporadic Alzheimer's Disease. This 3D miBrain-on-Chip could be harnessed to assess BBB penetration and downstream neurological effects and screen therapeutics towards optimizing delivery to the brain.

Fig. 1: Human iPSC-derived 3D Immuno-Glial-Neurovascular miBrain-on-Chip. **(A)** Schematic of miBrain formation harnessing patient-specific iPSCs differentiated into each of the resident brain cell types, encapsulated in Neuromatrix Hydrogel, and co-cultured for integral cell network self-assembly in our novel GelChip microfluidic platform to enable perfusable flow through vascular lumens within the miBrain, **(B)** distribution of (from left to right) iPSC-derived pericytes (cyan: mCherry-pericytes), astrocytes (cyan: mCherry-astrocytes), BMECs (cyan: mCherry-BMECs), neurons (cyan: tubulin neuron label), oligodendroglia (cyan: tdTomato pre-transfected oligodendroglia), and iMG (cyan: membrane pre-labeled iMG) throughout the full 3D miBrain, **(C)** neuronal phenotypes in miBrains cultured in dextran-based hydrogels fabricated with various brain ECM proteins (cyan: neurofilament, blue: Hoechst; scale bars, 50 μm), **(D)** example raster plots from MEA recordings of miBrains in Neuromatrix Hydrogel compared to Matrigel, **(E)** persistence of neurovascular unit phenotypes in miBrains cultured in Neuromatrix Hydrogel versus Matrigel after 5 weeks in culture (red: PECAM, cyan: TUBB3, blue: Hoechst; scale bar, 50 μm), **(F)** macroscopic view of gel structural integrity for miBrains cultured in VCN-incorporated engineered dextran-based hydrogel named Neuromatrix Hydrogel versus Matrigel after 4 weeks, **(G)** comparison of vascular networks in GelChips (right) compared to commercially available microfluidic platforms (left) (red: endothelial cells, scale bar, 200 μm); miBrains recapitulate key hallmarks of human brain tissue, inclusive of **(H)** BBB and neurovascular units with (*top*) 3D astrocytes distributed throughout BMEC networks throughout the miBrain (red: Imaris surfaces from ZO1-BMECs, green: mCherry-astrocytes; scale bar, 500 μm), (*bottom, left*) astrocytes with end-feet extending to vessels expressing canonical aquaporin-4 transporter (green: GFAP, gray: AQP4, red: PECAM, blue: Hoechst; scale bar, 30 μm; insert, gray: AQP4), and (*bottom, right*) anterior view of a miBrain 3D rendering (red: PECAM, cyan: neurofilament, blue: Hoechst; scale bar, 50 μm), (**G**) microglia: (*top*) 3D iMG distributed throughout BMEC networks throughout the miBrain (red: Imaris surfaces from mCherry-BMECs, green: Imaris surfaces from membrane pre-labeled iMG; scale bar, 500 μm), (*bottom, left*) visualized at higher magnification (red: Imaris surfaces from mCherry-BMECs, green: Imaris surfaces from membrane pre-labeled iMG; scale bar, 100 μm), and (*bottom, right*) 3D iMG distributed throughout neuronal networks throughout the miBrain (cyan: Imaris reconstruction of tubulin neuron label, green: Imaris surfaces from membrane pre-labeled iMG; scale bar, 100 μm), and (**H**) myelinated neuronal networks: (*top*) 3D oligodendroglia distributed throughout neuronal networks throughout the miBrain (cyan: tubulin neuron label, green: Imaris surfaces from tdTomato pre-transfected oligodendroglia; scale bar, 500 μm), (*bottom, left*) myelin dye-labeled neurons and oligodendroglia (red: FluoroMyelin, cyan: tubulin neuron label, green: tdTomato pre-transfected oligodendroglia, scale bar, 100 μm), and (*bottom, right*) visualized also via Imaris reconstructions for myelin and oligodendroglia (red: Imaris surfaces of FluoroMyelin, green: Imaris surfaces of tdTomato pre-transfected oligodendroglia; scale bar, 100 μm).

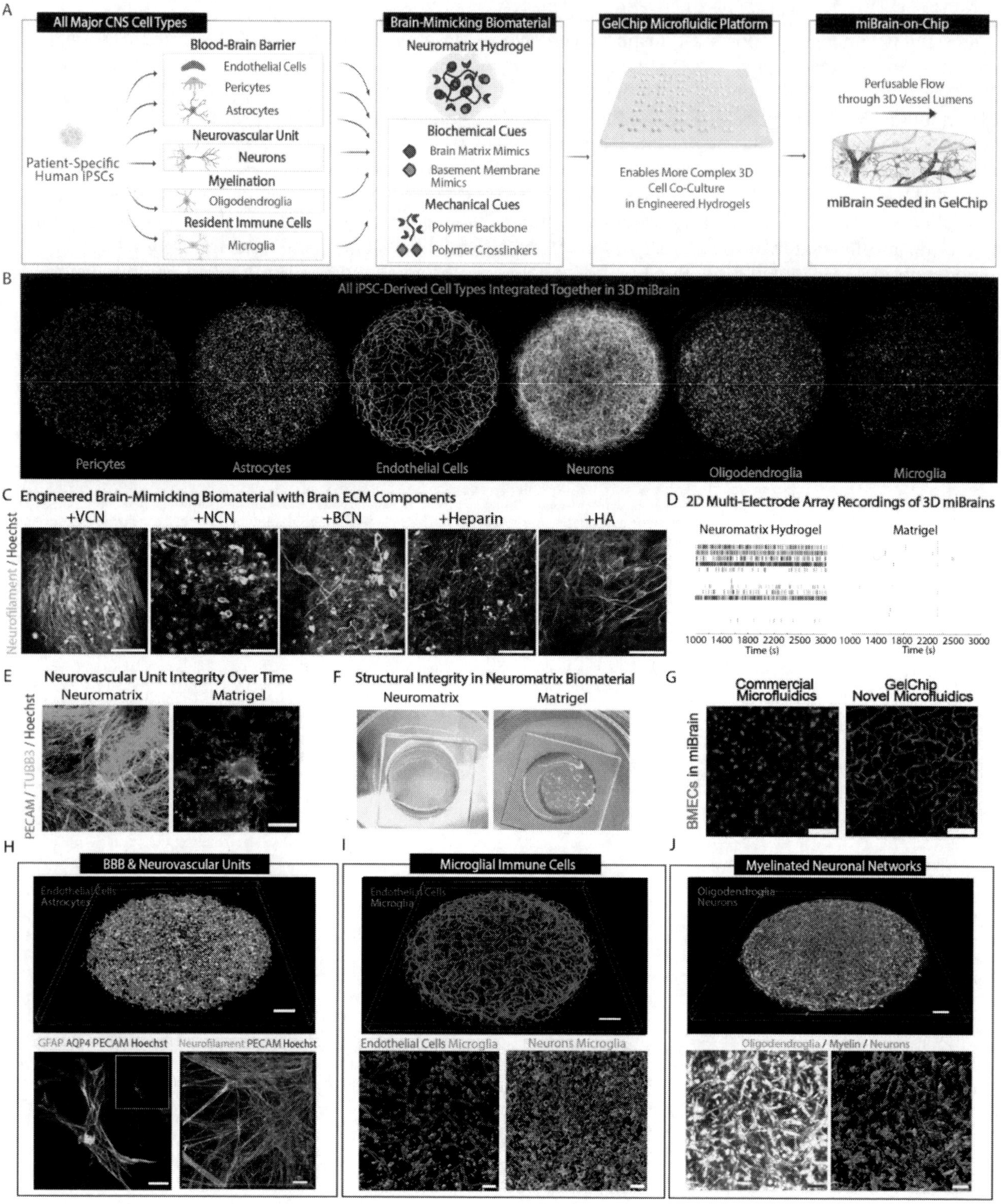
A
All Major CNS Cell Types
Blood-Brain Barrier
Endothelial Cells
Pericytes
Astrocytes
Neurovascular Unit
Neurons
Myelination
Oligodendroglia
Resident Immune Cells
Microglia
Patient-Specific Human iPSCs
Brain-Mimicking Biomaterial
Neuromatrix Hydrogel
Biochemical Cues
Brain Matrix Mimics
Basement Membrane Mimics
Mechanical Cues
Polymer Backbone
Polymer Crosslinkers
GelChip Microfluidic Platform
Enables More Complex 3D Cell Co-Culture in Engineered Hydrogels
miBrain-on-Chip
Perfusable Flow through 3D Vessel Lumens
miBrain Seeded in GelChip
B
All iPSC-Derived Cell Types Integrated Together in 3D miBrain
Pericytes
Astrocytes
Endothelial Cells
Neurons
Oligodendroglia
Microglia
C Engineered Brain-Mimicking Biomaterial with Brain ECM Components
Neurofilament / Hoechst
+VCN
+NCN
+BCN
+Heparin
+HA
D 2D Multi-Electrode Array Recordings of 3D miBrains
Neuromatrix Hydrogel
Matrigel
1000 1400 1800 2200 2500 3000
Time (s)
E
Neurovascular Unit Integrity Over Time
Neuromatrix
Matrigel
PECAM / TUBB3 / Hoechst
F
Structural Integrity in Neuromatrix Biomaterial
Neuromatrix
Matrigel
G
Commercial Microfluidics
GelChip Novel Microfluidics
BMECs in miBrain
H
BBB & Neurovascular Units
Endothelial Cells
Astrocytes
GFAP AQP4 PECAM Hoechst
Neurofilament PECAM Hoechst
I
Microglial Immune Cells
Endothelial Cells
Microglia
Endothelial Cells Microglia
Neurons Microglia
J
Myelinated Neuronal Networks
Oligodendroglia
Neurons
Oligodendroglia / Myelin / Neurons

355d Assessing Human Articular Cartilage Transcriptome Layers with RNA Sequencing

Pakkanpat Pondipornnont, Bernard Van Wie, Ryan Driscoll, Iwona Driskell, Lawrence J. Bonassar, Arda Gozen, Wenji Dong, Dominic Scalise, Mark Wildung, Salman Matan, Terreill Robertson

Osteoarthritis, a chronic disease, remains an issue for adults that causes cartilage degradation within a joint . According to the Centers for Disease Control and Prevention (2023), over 32.5 million adults in the US are affected by osteoarthritis (OA) [1]. In this study we seek to understand the connection between tissue engineering and genetics to regenerate human articular cartilage (hAC). We purpose to validate a protocol for RNA isolation and characterize the transcriptome of hAC in a tri-layer fashion via bulk RNA sequencing (bulk-RNA-seq). Additionally, we aim to analyze the transcriptome of normal articular cartilage in comparison to the hAC chemical composition and physical properties. We are relating these properties to the tri-layers of hAC through histological staining with Safranin O—Fast green and imaging with differential interference contrast (DIC) microscopy. We are relating these properties to superficial, middle, and deep zone with a cryotome procedure, RNA extracted, and qualifiedd by Bioanalyzer. Next, we generate bulk RNA sequencing of hAC layer-by-layer and compare results to early passaging of Mesenchymal Stromal Cells (MSC) and tissues intended for Matrix-Induced Autologous Chondrocyte Implantation (MACI). We will use differential gene expression (DE) analysis by DESeq2 R package software for bulk-RNA-seq. The result will be interpreted in terms of differentiation from MSCs to gene expression patterns of tri-layer hAC. We will report on development and validation of protocols for isolating cells and their subsequent characterization with application in regenerating the tri-layered hAC transcriptome stimulatory bioreactors used in our laboratory and corresponding properties of the extracellular matrix (ECM).

[1] “Osteoarthritis | CDC.” Accessed: Apr. 05, 2024. [Online]. Available: https://www.cdc.gov/arthritis/types/osteoarthritis.htm

355e Flexible Gallium-Based PEDOT:BF4 Coated Electrodes for Chronic Neural Recording and Biocompatibility

Alexandra Boyadzhiev, Huanan Zhang

Implantable microelectrode arrays are valuable tools in neuroscience for studying neural functionalities and disorders like Parkinson's, epilepsy, and spinal cord injury. However, rigid-metal-based electrode arrays often face challenges due to mechanical mismatches with brain tissue, leading to chronic inflammatory responses and compromised performance. To address this, we propose using gallium-based electrodes, offering minimal cytotoxicity, excellent electrical conductivity, and mechanical properties suitable for brain tissue. The low melting point of gallium allows it to transition into liquid after insertion, minimizing tissue-device mismatch.

Our study presents a flexible, gallium-based, PEDOT: BF4-coated neural array for chronic in vivo applications. Compared to rigid metal arrays, our gallium-based neuro devices demonstrate improved biocompatibility, as evidenced by reduced neuron death around the electrode implantation site in rat brains after eight weeks. We detail the fabrication process, including electrode fabrication, customizable array assembly, and PEDOT: BF4 deposition.

We characterize the electrochemical properties, thermal stability, and morphology of the PEDOT: BF4-coated gallium electrodes, showing their structural integrity and chronic electrochemical stability in vitro over five weeks. Acute in vitro and in vivo studies validate the action potential recording capabilities of the flexible gallium-based neuro device.

Furthermore, our chronic implantation study in rats showed that our devices maintained electrochemical functionality over eight weeks. Post-implantation analysis also reveals reduced inflammation GFAP markers and closer proximity of NeuN neuronal markers to the electrode site than rigid metal implants, suggesting enhanced biocompatibility and minimized tissue damage with our gallium-based, PEDOT: BF4-coated neural implant.

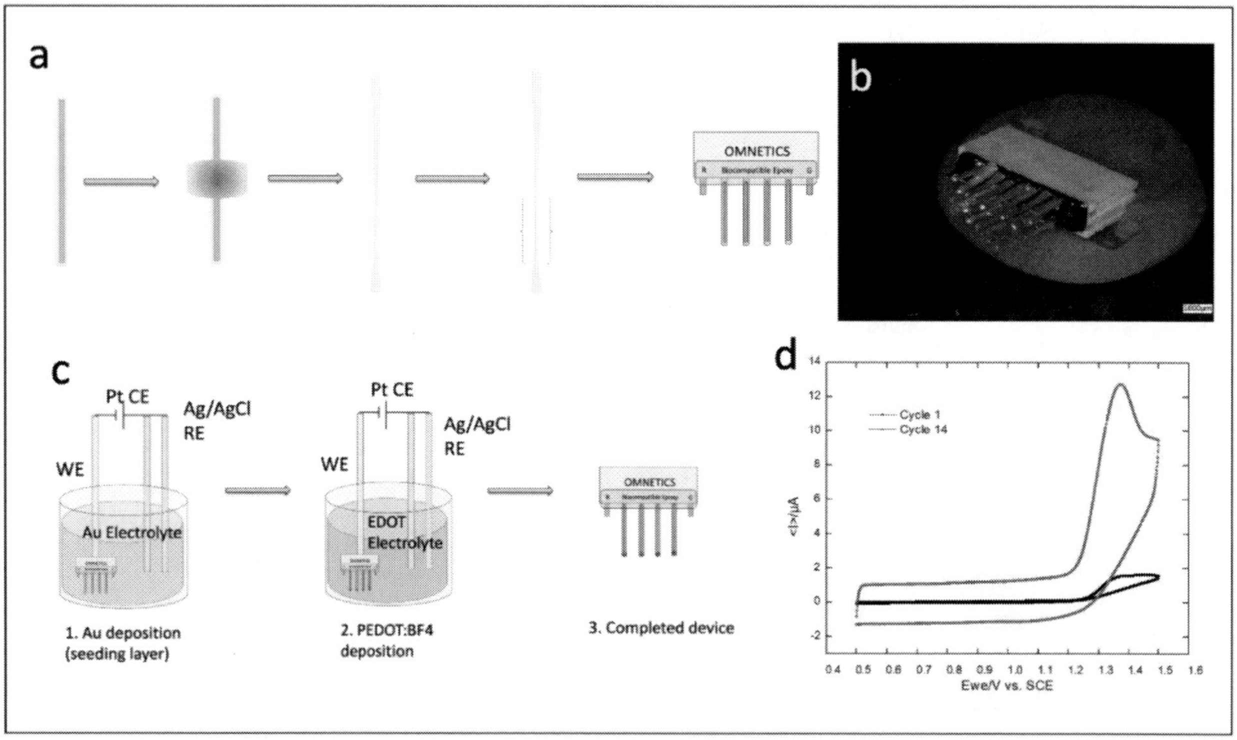

355f Robust (Dopamine)$_n$-Polycarboxybetaine Coatings for Blood-Contacting Devices Reduce Non-Specific Protein Adsorption and Improve Anticoagulation

Chenjue Tang, Di Liu, Suji Shin, Zuoming Wang, Mason Salonic, Lanyu Zheng, Caitlin Sigda, Keith Cook, Shaoyi Jiang

Polydopamine coating technology is a versatile and widely applicable surface modification method, inspired by the strong adhesion capabilities of mussels to rocks, attributing to the presence of abundant dopamine units in mussel adhesive proteins. Since 2007, this method has drawn significant attention due to its simplicity, multifunctionality, and broad applicability. Zwitterionic materials are a class of materials that contain both cationic and anionic groups but are still in neutral status and are known for their extraordinary anti-fouling properties. Therefore, a polymer, taking advantage of both moieties as described above, can be easily coated onto any substrate for anti-fouling purposes. In this work, we successfully combine features of both moieties and synthesize (dopamine)$_{1-4}$-polycarboxybetaine (DOPA$_{1-4}$-PCB, 10-40kDa) polymer, which can be easily utilized to modify any hydrophobic or hydrophilic surfaces. By performing an ELISA test with human fibrinogen on different polymer groups, it is found that anti-fouling performance depends on the molecular weight of PCB and the amount of dopamine moieties. Therefore, a polymer with an optimized structure with balanced two parameters will lead to excellent durability and antifouling properties.

With an optimized polymer, we coated the artificial lung devices attached to rabbits to avoid thrombosis and expand the lifespan of this ECMO circuit. The device failure time of the heparin+uncoated group was 3.31 hr, while the heparin+PCB-coated was 13.0 hr. The average resistance (mmHg/mL/min) of the heparin group was 1125, while the heparin+PCB-coated group was significantly lower at 385. In conclusion, these results demonstrate that the PCB coating significantly improves the anticoagulation of the ECMO circuit.

355g Next Generation Multi Modal Endoscopic Tattoo Inks for Precise Marking of Gastrointestinal (GI)-Lesions

Mallikarjun Gosangi, Harsh Sant, Vanshika Singh, Trishita Chowdhury, Jordan Yaron, Rahul Pannala, Kaushal Rege

Colorectal cancer (CRC) is the second leading cause of cancer related deaths in United States. In 2023, about 153,020 people will be diagnosed with (CRC) and 52,550 will die from the disease. As a result, early detection, and monitoring of suspicious lesions, including polyps, is critical to improving surgical outcomes and survival in patients. Colorectal lesions or polyps are marked using a tattoo ink to facilitate open abdominal or laparoscopic intervention to remove disease tissue or for long-term follow-up. However, the only currently available FDA-approved ink has significant drawbacks of which is loss of spatial retention, which is critical to the performance of the ink. Moreover, the ink suffers from fibrotic inflammation and peritonitis. Finally, a significant limitation is that these inks are not amenable to multimodal imaging, and therefore always require endoscopic intervention during longitudinal monitoring. There is a critical need for new tattoo inks that can improve surgical outcomes in colorectal endoscopy. Cationic chitosan derivative was used to encapsulate a library of anionic iron-oxide nanoparticles (IO based) and an anionic dual dye (indocyanine green (ICG) and rose bengal (RB)) (DD based) mixture to generate Tissue-Adhesive Tattoos generation-2 (TAT2) inks. These inks will stay with no diffusion to create a precise marking visualized under endoscopic white light (IO based and DD based) and laparoscopic near infrared light (NIR) (DD based).

The TAT2 inks were characterized using dynamic light scattering (DLS), transmission electronic microscopy (TEM), scanning electron microscopy (SEM), and viscosity to determine their size, morphology, shelf-life stability and injectability. The pro-inflammatory potential of both TAT2 library was assessed using the J774-DUAL mouse macrophage reporter cell line *in vitro*. IO based and DD based TAT2 candidates that demonstrated low inflammatory activity and high colloidal stability were evaluated *ex vivo* in porcine intestinal tissues and demonstrated excellent tight localization and similar contrast to Spot Ex (FDA approved), In addition, the NIR fluorescent potentials of DD based TAT2 were quantified using IVIS imaging of porcine colon tissue used for *ex vivo*. In vivo biocompatibility and performance of IO based TAT2 was evaluated by subcutaneous implantation in BALB/c mice, followed for 28-days replicated similar performance as observed in ex vivo and no adverse cell reactions. Endoscopic submucosal injection of four lead IO based inks were carried out in the colon of Yorkshire porcine and the efficacy and clinical toxicity were evaluated after 14-days. The *in vivo* performance of DD based TAT2 is under evaluation.

Both *in vivo* studies (mice and pigs) of lead IO based TAT2 candidates demonstrated superior spot localization performance while maintaining equivalent contrast to the commercial ink. Blood chemistry and histopathology of key organs of all the pigs used in the study indicated no toxicity of the TAT2 inks. The inks also demonstrated magnetic resonance contrast in post-mortem mice. All four novel inks outperformed the commercial inks when comparing spot localization with equivalent contrast performance in mice and pigs with variable tissue interactions.

We have developed an advanced set of next-generation endoscopic TAT2 inks that outperform existing options. These inks show improved precision in spot localization, exceptional contrast, minimal inflammatory effects, and the ability to be used for multimodal imaging. This innovative approach to clinical imaging holds promises for enhancing clinical outcomes in patients with gastrointestinal diseases.

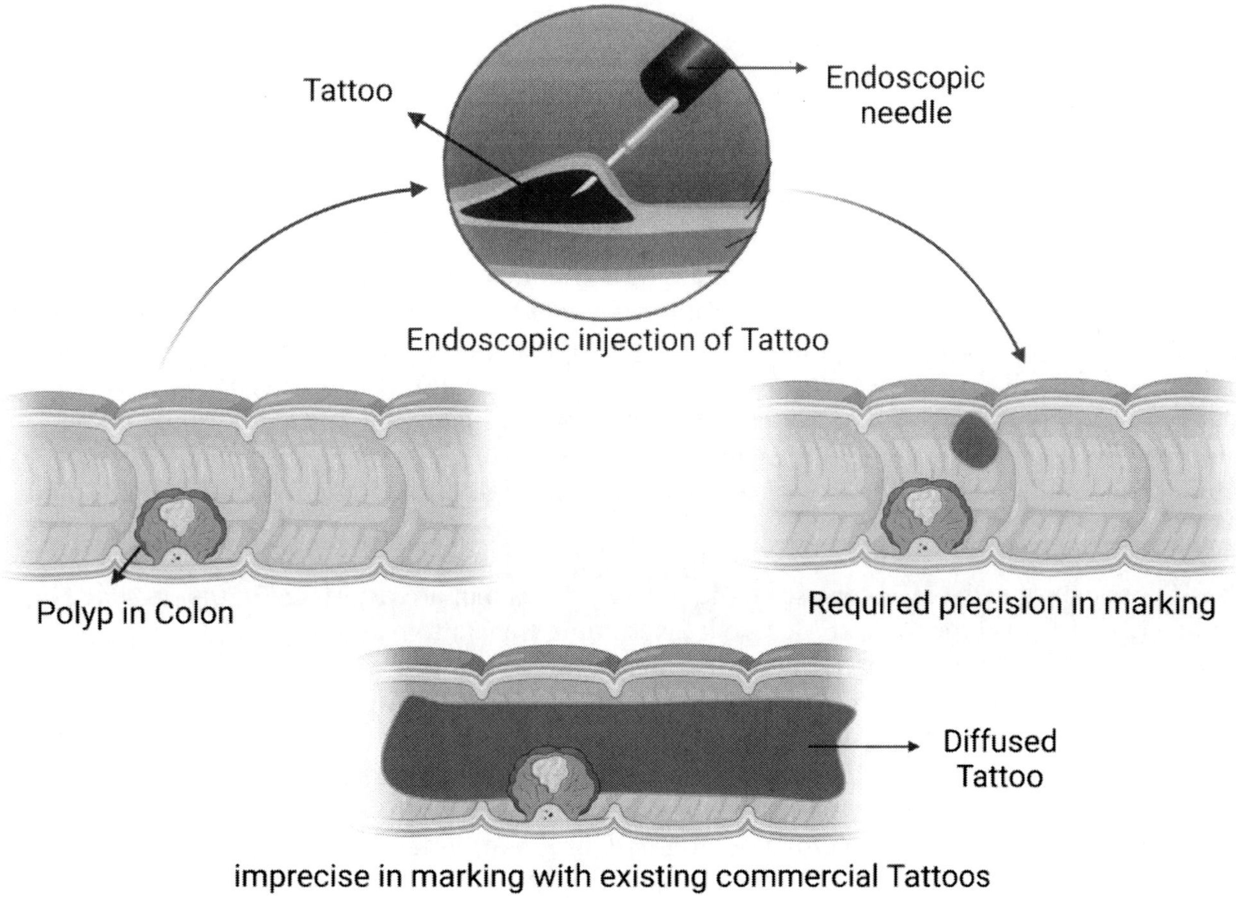

Tattoo
Endoscopic
needle
Endoscopic injection of Tattoo
Polyp in Colon
Required precision in marking
Diffused
Tattoo
imprecise in marking with existing commercial Tattoos

436a Deciphering the Mechanisms of Fluoroquinolone-Mediated Mutagenesis

Jenet Narzary, Sreyashi Ghosh, Mehmet Orman

Antimicrobial resistance continues to remain a critical global health issue. Unfortunately, there is an insufficient research and development pipeline, and more steps are urgently required to effectively implement better strategies in drug design and delivery. Bacteria need to protect their genomic material from numerous chemical mutagens in circulation to maintain genomic integrity. One common mutagen is the fluoroquinolone class of antibiotics, which are commonly used for the treatment of infectious diseases. Fluoroquinolones target DNA gyrase and topoisomerase IV enzymes during DNA synthesis, generating double-strand breaks in DNA, which activates the SOS regulatory network (1). The SOS response induces error-prone DNA repair mechanisms that can result in mutagenesis. A multitude of intricate elements that are yet mostly unknown contribute to the complexity of the mutagenic SOS response. Gaining a thorough grasp of this phenomenon could be highly beneficial in identifying new targets that can help fight against the formation of resistant pathogens.

The SOS response network comprises more than 50 genes that are expressed in the presence of DNA damage (2), and this process needs active transcription, translation, and ATP production. When we treated cells with ciprofloxacin, a fluoroquinolone antibiotic, along with metabolic inhibitors to inhibit transcription, translation, and ATP production during the recovery of *Escherichia coli* cells after the fluoroquinolone treatment, we were able to impair fluoroquinolone-mediated mutagenesis (3). As transcription and translation inhibition affects the downstream mechanisms of RecA, the master regulator of the SOS network, we focused on gaining further insights into these mechanisms by performing high-throughput screening of an *E. coli* promoter library, containing promoters of the major SOS response genes. We identified several candidates with upregulated expression; the *recA* gene exhibited the highest upregulation, followed by *recN, polB, dinB,* and *sulA,* among others. Deletion of certain SOS genes, either individually or in combination, resulted in an increased sensitivity towards ciprofloxacin along with a significant reduction in intracellular mutagenesis. Overall, this study highlights the potential downstream targets of RecA that can be important in understanding and controlling intracellular mutagenesis in the context of antibiotic resistance.

References

1. Hooper, D. C. (1999). Mechanisms of fluoroquinolone resistance. Drug Resistance Updates, 2(1), 38–55. https://doi.org/https://doi.org/10.1054/drup.1998.0068

2. Maslowska, K. H., Makiela-Dzbenska, K., & Fijalkowska, I. J. (2019). The SOS system: A complex and tightly regulated response to DNA damage. Environmental and molecular mutagenesis, 60(4), 368–384. https://doi.org/10.1002/em.22267

3. Sreyashi, G., & A, O. M. (2024). Exploring the links between SOS response, mutagenesis, and resistance during the recovery period. Antimicrobial Agents and Chemotherapy, 0(0), e01462-23. https://doi.org/10.1128/aac.01462-23

436b Accelerating Pathogen Detection: Towards Portable, Rapid Isothermal PCR Diagnostics

MinGin Kim, Vijay Ravisankar, Yassin Hassan, Victor M. Ugaz

The polymerase chain reaction (PCR) is a gold standard for nucleic acid-based diagnostics, but portability and accessibility are constrained by the electrical power associated with cycling through different temperatures to denature, anneal, and extend the DNA strands. This constraint significantly limits the portability and accessibility of PCR, especially critical in pandemic response and public health monitoring. Consequently, the ability to perform PCR isothermally has long been sought-after to overcome these barriers. However, the PCR-like isothermal DNA replication methods developed thus far (e.g., loop-mediated isothermal amplification (LAMP)), while promising in specific applications, have yet to consistently match the performance and reliability of PCR. Here, we report a discovery that overcomes these limitations by enabling rapid isothermal PCR to be achieved with 100% repeatability. This breakthrough is accomplished by manipulating the interplay between the PCR biochemistry (via the amplicon GC content) and the lava-lamp-like microscale flow established inside a PCR tube whose upper and lower surfaces are isothermally maintained at annealing and denaturing temperatures. Surprisingly, we find that optimal performance critically depends on selecting primers that replicate amplicons with high GC content, contradicting established PCR primer design rules. Our approach accelerates PCR to speeds that rival ultra-fast instruments, enabling rapid, repeatable isothermal PCR across various targets relevant to diagnostics and pathogen detection. By allowing rapid repeatable isothermal PCR, our findings pave the way to make nucleic acid-based analysis systems accessible, reliable, and efficient so that lab-quality diagnostics can be broadly and equitably deployed.

436d Sars-Cov-2 Real-Time Airborne Transmission within Different Departments of a Hospital COVID Building, and Evaluation of Air Cleaners in Reduction of Viral Load.

Ilias Frydas, Marianthi Kermenidou, Maria Karypidou, Spyros Karakitsios, Kostas Stamatopoulos, Dimosthenis Sarigiannis

The pandemic of COVID-19 has brought in light the necessity for the development of novel detection methods for airborne transmitted pathogens on the one hand and the importance of effective measures in hospital departments on the other. In this study, Sars-CoV-2 detection has been performed from December 2021 till May 2022 in different departments of the COVID building at the Ippokrateio University Hospital in Thessaloniki, Greece. More specifically, cascade impactors and air pumps with filters were placed at the first floor in the ICU (Intensive Care Unit), the HDU (High-Dependency Unit), in personnel restrooms, in the cloth changing rooms, and in corridors and rooms at the at the COVID second floor. Sampling was performed every 24 or 48 hours till seven days continuously. In addition, in a COVID room with confirmed viral load an air cleaner machine was placed to examine the effect on viral load in hospital environment. Air filters were subjected in RNA extraction using an in-house developed method, and RT-qPCR was performed to detect and quantify viral copies/m^3. Results showed that at the ICU and the HDU departments in which negative pressure air filtration with HEPA filters is applied no viral copies were detected (Table 1). On the contrary, viral load was effectively detected at the COVID floor in rooms and corridors and ranged from 25,9 to 1123,7 copies/m^3. In addition, viral load in the COVID room where an air cleaner was placed showed a reduction of 95,7%. Furthermore, sequencing of SARS-CoV-2 positive samples from January, February and May 2022 showed that during our study Omicron (B.1.1.529) was the dominant variant. In conclusion, in this study SARS-CoV-2 was effectively detected in real-time after extended sampling and ranged between 24 h to 7 days, and it was shown how important and effective are air cleaners as first-line measures against the pathogen airborne transmission in hospital environments.

Table 1. Overview of air sampling and virus detection in the different departments and floors of the Ippokrateio Hospital COVID building. Sampling duration is presented in minutes and volume in air liters per minute (L/min). Real-time quantitative polymerase chain reaction (RT-qPCR) has targeted S and ORF1ab genes of SARS-CoV-2 and results are shown in copies/m^3.

Hospital Areas	Sampling start date	Volume (L/min)	Sampling duration (min)	Total Volume (L)	RT-PCR	S gene copies/m^3	ORF1ab copies/m^3
ICU (1st floor)							
	21/12/2021	9	4310	38790	Negative	0	0
	24/12/2021	9	2855	25695	Negative	0	0
	26/12/2021	9	4290	38610	Negative	0	0
	29/12/2021	9	7217	64953	Negative	0	0
	03/01/2022	9	2875	25875	Negative	0	0
	05/01/2022	9	2875	25875	Negative	0	0
	07/01/2022	9	5715	51435	Negative	0	0
	11/01/2022	9	1425	12825	Negative	0	0
HDU (1st floor)							
	14/01/2022	9	4295	38655	Negative	0	0
	17/01/2022	9	2895	26055	Negative	0	0
	19/01/2022	9	2895	26055	Negative	0	0
	24/01/2022	9	4125	37125	Negative	0	0
ICU Drug preparation room (1st floor)							
	18/03/2022	9	4320	38880	Negative	0	0
	21/03/2022	9	2880	25920	Negative	0	0
ICU Personel restroom (1st floor)							
	29/03/2022	9	4050	36450	Positive	8,4	3,3
	01/04/2022	9	7530	67770	Positive	5,1	4,6
	06/04/2022	9	2715	24435	Negative	0	0
COVID Clinic Room (2nd floor)							
	27/01/2022	9	1440	12960	Positive	679,2	586,4
	31/01/2022	9	2865	25785	Positive	885,3	323,1
	[illegible]	[illegible]	[illegible]	[illegible]	[illegible]	[illegible]	[illegible]

	02/02/2022	9	3035	27315	Positive	1123,7	876,6
	04/02/2022	9	4300	38700	Positive	964,4	514,7
COVID Clinic Corridor (2nd floor)							
	07/02/2022	9	2855	25695	Positive	211,6	116,1
	09/02/2022	9	2880	25920	Positive	308,3	157,2
	11/02/2022	9	4335	39015	Positive	220,8	195,6
	14/02/2022	9	2865	25785	Positive	403,7	226,8
	16/02/2022	9	2825	25425	Positive	379,5	310
	18/02/2022	9	4320	38880	Positive	481,8	277,3
COVID Clinic Personel Restroom (2nd floor)							
	21/02/2022	9	10080	90720	Positive	156,1	68,5
	28/02/2022	9	2880	25920	Positive	101,4	51,8
	02/03/2022	9	2880	25920	Positive	92,7	25,9
COVID Clinic Consumables storage room (2nd floor)							
	04/03/2022	9	5790	52110	Positive	133,4	118,2
	08/03/2022	9	4350	39150	Positive	76,1	49,6
COVID Clinic Room 204 without Aircleaner (2nd floor)							
	01/05/2022	9	2850	25650	Positive	550,5	420,1
	03/05/2022	9	5680	51120	Positive	159,5	136,4
COVID Clinic Room 204 with Aircleaner (2nd floor)							
	06/05/2022	9	5640	50760	Positive	3,4	5,3
	10/05/2022	9	3000	27000	Positive	10,4	5,2

436e Predicting the Impact of Influenza A Virus Mutations on the Display of Viral Fragments to Helper T Cells

Mercedes Haley, James Gillespie, Robert J. Pantazes

A virus can mutate its genetic code through either antigenic shift or antigenic drift, which can cause changes to the protein that the genes encode. Mutations in the protein canimprove functionality and survivability, which later variants maintain. Additionally, viral mutations can affect how a person's immune system detects a mutated virus. Immune cells, such as T cells and antibodies, created during previous infections do not have the same detection capability when the virus protein is mutated. For this study, the change in immune recognition of helper T cells in humans was predicted for mutated Influenza A virus strains.

The 2009 Influenza A virus rapidly spread, causing the 2009 H1N1 pandemic. This virus is known as a zoonotic virus since it was transmitted from swine to humans. The primary Influenza proteins detected by the immune system are the surface proteins, Hemagglutinin (HA) and Neuraminidase (NA). Helper T cells detect small peptides of the virus that have been broken down and displayed on the surface of professional antigen-presenting cells (APC), such as macrophages. The peptides displayed to helper T cells are chosen based on their compatibility with the binding grooves of Major Histocompatibility Complex (MHC) class II molecules. Many different alleles encode these molecules, which have been passed down genetically. Each MHC Class II allele has different affinities for a variety of peptides. Thus, changes in the protein sequence can either improve or decline the presentation of peptides to the helper T cells. Vaccine development for influenza is difficult due to the mutation's speed and the different capabilities of detecting the virus within different ethnicities. Thus, getting a better insight into the effects of mutations in the Influenza virus on the MHC class II presentation to T cells could provide important insight.

The consensus sequences for every month during the 2009 pandemic needed to be identified for the two surface proteins. With the number of sequences available on the Influenza Virus Database provided by NCBI, the sequences were separated by month and submitted to ClustalOmega, a Multiple Sequence Alignment (MSA) software. With the sequences for each month aligned, a consensus sequence could be identified by designating each position with the highest percentage of amino acid for that position.

With access to the consensus sequences from April 2009 to October 2010, the change in helper T cell immune recognition could be calculated by comparing two consecutive sequences, with the first month as the wild type and the second month designated as the mutated sequence. Predicting the binding scores between the MHC class II molecules and the viral peptides can be done with ProPred, which can predict the binding scores of 51 alleles. The binding score between the wildtype and mutated sequences was subtracted and then divided over the highest possible binding score for that allele to understand the change in immune recognition.

While this does give the change in binding score, it does not provide a concise and meaningful result. Thus, a weighted score was determined using the allele frequencies for each ethnicity, obtained from the Allele Frequency Net Database, compared to the change in binding score for each allele. The products of each allele binding score and frequency were then added to create the weighted change in binding for each ethnicity.

After transitioning to humans, the 2009 Influenza A Virus showed increased immune recognition in the first couple of months. This is thought to be due to the virus improving its survivability. In the later part of the pandemic, the accumulated mutations decreased immune recognition, promoting immune evasion. When looking at the impacts on different ethnicities, the general trend was for all the weighted scores for the ethnicities to increase if the immune recognition increased, but at various amounts. This was the same for if the mutations caused a decrease in immune recognition. The consensus sequences for each month after the 2009 Influenza pandemic will be analyzed for further study. This will include investigating the different subtypes that have circulated the world, such as H3N2. It is also of interest to compare regional strains to those used to create the vaccine to understand the efficacy of designed vaccines.

436f Systemic RNA Delivery By Phosphatidylserine Lipid Nanoparticles for Efficient Immunotherapy

Sijin Luozhong, Ruoxin Li, Kay McIlhenny, Yuping Hu, Shaoyi Jiang

The development of mRNA-lipid nanoparticles (mRNA-LNP) marks a revolutionary advance in medical science by enabling diverse therapeutic applications, including cancer vaccines and immune cell engineering, through targeted and efficient mRNA delivery to specific cells without adverse immune responses. Systemic RNA delivery to the secondary lymphoid organs (SLOs) for immunotherapy offers distinct advantages over traditional intramuscular methods, including an increased amount of antigen administration and direct access to most immune cells and lymphoid organs in the circulatory system. Current strategies of SLO-targeted delivery primarily rely on a passive mechanism that involves introducing an excessive amount of anionic charge to the formulation. However, this method comes at the cost of reduced transfection efficiency. Here, we report a new LNP system incorporating phosphatidylserine (PS) for the efficient systemic delivery of mRNA to SLOs. PS is known to bind endocytosis receptors on macrophages and is used by many viruses for enhanced cellular uptake. Our initial result demonstrated that the PS-LNP performed efficient mRNA expression in the spleen and lymph nodes. Mechanistic studies have verified its SLOs-targeting effect is mediated by monocytes/macrophages. A Design of Experiment (DOE) approach was further implemented to optimize the PS-LNP for improved mRNA transfection. The optimized formulation was validated in primary macrophages and in vivo, exhibiting more potent luciferase-mRNA expression than commercial formulations. Furthermore, it was validated to perform macrophage-specific expression by Cre-mediated editing as well as Chimeric-antigen engineering experiment, indicating promising in vivo macrophage engineering applications. Additionally, it induced potent antigen-specific T-cell response upon systemic vaccination, suggesting its viability as a promising platform for systemic immunotherapy.

436g Enhancing mRNA Vaccine Immunogenicity By Engineering the Lipid Nanoparticle and mRNA-Encoded Antigen

Allen Jiang, Bowen LI, Idris Raji, Daniel G. Anderson, Robert Langer

mRNA vaccines have demonstrated great success against SARS-CoV-2. However, further clinical translation of mRNA vaccines will require eliciting optimal immune responses with desirable safety profiles. To further improve the safety and efficacy of mRNA vaccines, we here engineered a multiply self-adjuvanted mRNA vaccine system whereby both the ionizable lipid as well as the mRNA-encoded antigen are designed to potentiate the elicited immune response. We hypothesized that incorporating multiple self-adjuvanting properties directly into the mRNA-LNP formulation would spatially and temporally restrict immunostimulation to sites of antigen presence, improving beneficial, antigen-specific immune responses while limiting detrimental systemic inflammation.

We first screened a combinatorial library of 480 biodegradable ionizable lipids with adjuvanting cyclic amine headgroups, identifying a top ionizable lipid with enhanced innate immunostimulatory properties leading to a potentiated immune response when compared to the FDA-approved ionizable lipid MC3. We then designed an adjuvanted mRNA transcript encoding SARS-CoV-2 model antigens genetically fused to C3d, a terminal degradation product of complement, as a molecular adjuvant. Genetic fusion of C3d to the encoded antigen targets the fusion protein to complement receptor 2 on B-cells and follicular dendritic cells, lowering the barrier to activation of these cells which play central roles in coordinating the innate and adaptive immune responses. Compared to the mRNA encoding viral antigens alone, the inclusion of C3d in the mRNA transcript increased the magnitude of antigen-specific antibody titers by at least ten-fold in mouse sera. Notably, the enhanced immune response elicited by C3d did not elevate pro-inflammatory cytokines systemically, suggesting a favorable safety profile for this adjuvanting strategy. In mice, intramuscular and intranasal vaccination with LNPs combining the adjuvanted ionizable lipid and mRNA transcript resulted in synergistically improved immune responses to a SARS-CoV-2 Delta variant antigen when compared to unadjuvanted vaccines. Our results demonstrate that a multiply adjuvanted system holds great potential in improving the efficacy, safety, and ease of administration of mRNA vaccines.

436h In Vitro Transcription Platform Process Modelling to Ensure mRNA Vaccine Quality

Simon Daniel, Cleo Kontoravdi, Nilay Shah

mRNA is emerging as one of the most promising technologies in the medical field. Beyond COVID-19 vaccines, these new products are raising high hopes for new clinical application, including many unmet medical needs. Over a hundred new RNA-based vaccines have now entered the clinical pipeline. Strikingly, this technology offers the potential for modular, disease-agnostic manufacturing platforms. However, current production processes still present a high level of impurities and by-products in comparison to conventional vaccine technologies. Double-stranded RNA (dsRNA) and small RNA are considered particularly critical, as they are highly immunogenic. Disease-specific requirements, such as higher dose regimen or lower immunogenicity, could potentially prevent clinical translations. These products are also heterogenous in size and formation mechanisms, posing additional analytical challenges and downstream burdens. Furthermore, when switching from a sequence to another, significant variations in the RNA active substance quality profile have been reported. A deper process understanding appears thus essential to further optimize In Vitro Transcription (IvT) and ensure consistent product quality.

This study shows how process modelling tools can be used to enhance this knowledge and unlock RNA platform potential.

In line with Quality by Design (QbD) principles, a holistic modelling approach was implemented for multi-product IvT. Over 400 reactions and 30 RNA constructs were used to calibrate this platform process model. A system of differential algebraic equations (DAEs) was implemented and solved in Pyomo 5.7 (Python). The dynamics of more than 30 chemical species in solutions are thus captured under partial-equilibrium assumptions. This modelling framework represents the mechanistic backbone for the prediction of additional critical quality attributes (CQAs), including the formation of dsRNA and small RNAs. Then, an analytical workflow based on RNA capillary gel electrophoresis, ELISA, RP-IP-HPLC and spectrum analysis was developed to assess product quality profile at different sampling times. In addition, a screening design was implemented to identify the non-linear effects of plasmid DNA, T7 RNA polymerase (T7RNAP), magnesium, nucleotides and buffer concentrations for 12 novel RNA products. Purified mRNA molecules at different concentrations were also incubated without plasmid DNA and under different conditions to isolate templated from non-templated RNA degradation and by-products formation.

This work provides new insights into mRNA quality and a more complete view of the multiple species that make up the final IvT product. Model predictions were validated through cross validation algorithms. Such a mechanistic approach guarantees extrapolation capability with future RNA vaccine candidates and process innovations. Additionally, a multi-objective design space optimization workflow is presented. Operating regions meeting quality requirements are showcased and highlight the central role of T7RNAP activity. Eventually, a model-based strategy is introduced to effectively explore the IvT design space for regulatory purposes. The integration of a maximal covering design can provide a process validation tool, ensuring greater process flexibility and confidence within the normal operating regions.

Overall, this study provides a new model-based approach for IvT process optimization, capturing the kinetics of the formation of the most critical RNA quality attributes. It is also a step towards the deployment of versatile RNA production platforms. In the near future, the integration of sequence effects into this modelling framework could bridge the gap between process development and sequence design. It can also lay the foundations for new feeding strategies to ensure better control over process conditions, further guaranteeing the safety and efficacy of any new mRNA candidates.

502a Machine Learning-Based Method to Analyze Metabolic Fluxes of Patient Tumors

Baharan Meghdadi, Anjali Mittal, Andrew Scott, Sravya Palavalasa, Eric Peterson, Abhinav Achreja, Costas Lyssiotis, Wajd Al-Holou, Daniel Wahl, Deepak Nagrath

Personalized oncology must overcome the complex challenge of distinct cancer phenotypes in each patient, formed in part in response to unique metabolic crosstalk among cells in the tumor microenvironment which can be a determining factor in therapeutic response. Although isotope tracing sheds light on differences of metabolite enrichments in tumor and normal tissues, quantifying the metabolic fluxes in patients remains elusive. The conventional metabolic flux analysis tools are unable to estimate patient metabolic fluxes because we are limited to a single time point enrichment datapoint from each patient's in-vivo measurement at isotopic non-steady state. We overcame this challenge by generating training data from the limited patient data and implementing a machine learning model to predict fluxes. To generate training data, we first simulated constrained random fluxes and enrichment patterns at different time points using patient circulating metabolites enrichment via an isotopic non-stationary metabolic flux analysis framework. Using this data, we trained a convolutional neural network to predict flux ratios. To validate our model, we compared predicted flux ratios to the experimental data from mouse models for two different treatments. Our recent findings from a glioblastoma (GBM) mouse model study using conventional metabolic flux analysis revealed higher de novo guanosine monophosphate (GMP) synthesis in GBM than normal cortex. These tumors can be sensitive to mycophenolate mofetil (MMF) which inhibits inosine monophosphate dehydrogenase, but only if the major source of GMP production is de novo synthesis and not salvage. Our machine learning model can discriminate between sources of GMP synthesis in patients, thus enabling prediction of potential responders. Furthermore, we previously showed that serine uptake is higher in GBM than normal cortex, and a serine- and glycine-restricted diet slows tumor growth in mice. While both circulating serine and microenvironment-derived serine could account for serine uptake in GBM, we hypothesized that the circulating serine dependence would better predict response to dietary serine depletion. Hence, we combined patient scRNA-seq and metabolomics data to feed our neural network framework to estimate microenvironmental and circulating serine uptake in six patients with brain cancer. Thus, we have shown that our model can identify patients who will benefit from MMF treatment or a serine- and glycine-restricted diet, enabling administration of highly effective personalized treatments for GBM patients.

502b Dendritic Cell-Derived Endogenous Virus-like Vesicles As Cancer Vaccines

Wenchao Gu, Ruoxin Li, Hayleigh Goodrich, Dorin Artzi, Yi-Chih Tsai, Simian Cai, Ann Li, Aixin Shi, Vaibhav Upadhayay, Nadine Elkasri, Sijin Luozhong, Evelyn Goldwasser, Justin Lau, Ella Sultan, Qiuming Yu, Shaoyi Jiang

The quest for enhancing the potency of cancer immunotherapies has led to the exploration of dendritic cell-derived extracellular vesicles (DEVs) based vaccines, which, however, are hampered by limited T cell stimulatory capacities in recent clinical trials. Addressing this challenge, we incorporated endogenous virus-like components into DEVs to boost their T cell engagement. Our approach aims to enhance the effectiveness of T cell mediated tumor targeting and eradication, utilizing engineered DEVs tailored to mimic viral attributes.

To achieve this goal, we genetically modified dendritic cells (DCs), introducing virus-like components to create engineered virus-like vesicles (eVLVs). We first analyzed their composition with mass spectrometry and tracked their enrichment in the spleen and direct interaction with T cells using in vivo and confocal imaging. Next, using the Ovalbumin (Ova) model antigen, we demonstrated that administration of eVLVs presented Ova to T cells and enhanced robust and balanced immune responses. RT-qPCR analysis showed elevated immune marker levels, and flow cytometry identified increased activation of both CD4+ helper and CD8+ cytotoxic T cells. Memory T cell formation, verified by ELiSpot assays for interferon-gamma (IFNg), highlighted the eVLV's capacity for inducing long-term immunological memory. IgG antibody levels in the eVLV group showed a dramatic 20-fold increase over the experimental control after four weekly doses.

To assess eVLV's potential in cancer prevention, we generated a murine melanoma model by intradermally injecting B16-F10 melanoma cells into C57BL/6 mice. We monitored tumor growth via caliper measurements and computed tomography scans. Immune response evaluation employed flow cytometry to identify and quantify T cell subsets and other immune populations, ELISpot assays to measure IFNg secreting cells, and multiplex cytokine assays for systemic immune response profiling. The long-term efficacy and immune memory were tested by monitoring survival rates and performing a melanoma rechallenge in surviving mice. As a result, eVLV administration resulted in a 60% cure rate, with 67% of treated mice surviving more than 100 days post-melanoma rechallenge, markedly outperforming the control group, which had no survivors. This result emphasizes the potential of eVLVs for achieving lasting melanoma immunity and enhancing survival rates. Finally, our investigation into the molecular underpinnings of these effects uncovered a notable similarity between the eVLV surface proteins and retroviral envelope proteins, suggesting a new method for engaging T cell receptors.

In summary, our findings underscore the potential of utilizing endogenous virus-like properties of engineered DEVs to develop effective cancer vaccines. By harnessing the unique capabilities of eVLVs to recruit enveloping proteins and mimic viral engagement with immune cells, we present a new strategy for targeted cancer immunotherapy. This strategy also steers immunotherapy towards highly personalized treatments. Utilizing patient-derived donor DCs for vaccine production enhances personalization, ensuring vaccines are highly biocompatible and free from toxicity. This integration of biomedical engineering and immunology paves the way for tailored and targeted vaccines, offering a leap forward in the precision and safety of cancer immunotherapy.

502c DEEP Phenotyping of Immune CELLS to Distinguish Disease States in EARLY-STAGE Breast Cancer

Hunaiz Bin Navaid, Adity Pore, Siva A. Vanapalli

Background: Liquid biopsy, a non-invasive prognostic tool for monitoring disease states, holds significant promise for personalized and precision medicine in breast cancer management. Conventionally, it has concentrated on investigating various cancer-associated cells using specific biomarkers, with current research predominantly centered around the study of rare circulating tumor cells (CTCs). Here, we propose redirecting focus towards analyzing immune cell phenotypes, which are abundant and express several biomarkers. Our aim is to build a deep phenotyping framework capable of distinguishing disease states using immune cell biomarkers from individuals enabling personalized disease monitoring.

Methodology: To achieve our goal, we developed a computer vision framework to detect immune cells expressing the DAPI, CD45, and Vimentin biomarkers from immunostained images of peripheral blood of early-stage breast cancer patients. We extracted phenotypic features comprising immune cell morphology and biomarker expression, and used these multi-dimensional datasets to distinguish healthy and diseased patients. Datasets comprising 19 blood samples, 800,000 immune cells and 9 million phenotypic data points were used to conduct this study.

Results: Using a lower-dimensional representation of the extracted phenotypic features, we find clear distinction in data clusters associated with healthy, stage II and stage III breast cancer patients. To understand the phenotypic features that cause this distinction, we conducted statistical analysis between patient cohorts by comparing the variations observed with each biomarker. Our investigation highlighted the most notable differences in the nucleus of immune cells from diseased patients compared to those from healthy individuals, with morphological and biomarker expression features showing significant disparities [50% decrease & 100% increase in size & biomarker expression, respectively, $p < 0.05$]. Leveraging these insights, we quantitively predicted the disease states, with an accuracy > 80%, based on the phenotypic nature of the cells using logistic regression.

Conclusion: Our study reports a phenotypic framework to characterize immune cell phenotypes present in the liquid biopsy of early-stage breast cancer. We believe this innovative approach broadens the scope for clinical utility of liquid biopsy for precision medicine, paving the way for effective targeted therapeutic interventions in non-invasive, blood-based diagnostics.

502d Co-Stimulatory Molecules Increase the Immunogenicity of Tumor-Specific Neoantigens

Amy Laflin, Estefani Quinones, Yiwei Feng, Erica Wagner, Rachel Hercek, Angelica Bernal Penaloza, Ronit Kumar, Isha Arora, Nathan Brown, Shaoyi Jiang

The advancement of next-generation sequencing technologies has led to the identification of patient-specific tumor antigens. These tumor-specific antigens, termed neoantigens, arise from genomic mutations and can be used to create personalized cancer immunotherapies. In particular, neoantigen based vaccines are clinically being studied as a means to increase and diversify the anti-tumor T-cell responses in patients. Sequencing has allowed for large libraries of neoantigen sequences to be identified; however, only a small population of neoantigens are able to mount strong immune responses *in vivo*. This greatly limits the efficacy of personalized neoantigen based vaccines. There is a clinical need to increase the anti-tumor T-cell responses, or immunogenicity, that lowly immunogenic neoantigens produce. The goal of this study was to determine if co-stimulatory molecules could increase the immunogenicity of a low immunogenic neoantigen. Co-stimulatory molecules provide a secondary signal to T-cells and are required for full T-cell activation. We identified neoantigens in the murine Pan02 tumor model and ranked their immunogenicity using an enzyme-linked immunospot (ELISPOT) assay. A low immunogenic neoantigen, called Neoantigen 77, was unable to produce a significant immune response and was chosen as the model neoantigen for completion of this work.

We set up a screen of six co-stimulatory molecules predicted in literature to increase the Th1 cellular immune response when expressed on antigen presenting cells. Each individual co-stimulatory molecule and every pairing combination of molecules were co-delivered with Neoantigen 77 and the neoantigen-specific T-cell response determined. Specifically, mice were vaccinated intramuscularly with lipid nanoparticles (LNPs) co-delivering mRNA encoding Neoantigen 77 and encoding co-stimulatory molecules. Sixteen days after the first vaccine dose, splenocytes were isolated from mice and analyzed in an IFNy, IL-4 ELISPOT. ELISPOT data revealed that three of the individual co-stimulatory molecules significantly increased the number of activated anti-neoantigen 77 T-cells. Additionally, after screening every pairing combination of co-stimulatory molecules, it was found that several significantly increased the immunogenicity of Neoantigen 77. One pairing in particular had a 3.5 fold greater number of IFNy secreting Neoantigen 77 specific T-cells compared to mice injected with Neoantigen 77 alone. The optimal co-stimulatory pairing was dose optimized for further studies. Splenocytes were isolated from mice treated with Neoantigen 77 plus the optimized pairing, restimulated with dendritic cells presenting Neoantigen 77, and then stained for flow cytometry. Flow analysis revealed mice receiving the optimized co-stimulatory molecules had an increase in both IFNy secreting CD4+ and CD8+ T-cells, as well as higher ratios of CD8+ to regulatory T-cells (compared to control mice receiving Neoantigen 77 alone). Toxicity studies revealed that cytotoxicity remained low even after repeated doses of the optimal co-stimulatory pairing. Mouse body weight was measured, and all biological replicates remained within 10% of starting weight, even after four injections of co-stimulatory molecules given over a two-week period. Liver enzymes, AST and ALT, did not increase after vaccination, indicating low liver toxicity. Additionally, H&E staining of liver, lungs, kidney, and brain of mice treated with the optimized dose of co-stimulatory molecules did not show any significant organ toxicity.

In conclusion, this is the first study to our knowledge that has tested co-stimulatory molecules in the context of a mRNA-LNP cancer vaccine. It was shown that co-stimulatory molecules can increase the immunogenicity of a low immunogenic neoantigen while maintaining low toxicity. Future work for this research will test a neoantigen mRNA-LNP vaccine with the optimized co-stimulatory pair in the Pan02 pancreatic mouse model.

502e Zwitterionic 3D Culture Matrix Reduces Symbolic Gene Differential Expression in Prostate Cancer Organoids

Chenjue Tang, Di Liu, Erica Wagner, Wenchao Gu, Haoxuan He, Sizhe Huang, Shaoyi Jiang

Neuroendocrine prostate cancer (CRPC-NEPC) is one of the most aggressive subtypes among all prostate cancer subtypes, and currently there are no target therapies for this disease available. A major hindrance to therapy development is that there is not an accurate disease model available for researchers and clinicians to understand the oncologic transformation mechanism that results in clinical heterogeneity. Matrigel is a naturally derived material that has been shown to promote organoid growth and self-assembly. However, its xenogeneic source and batch-to-batch variation severely limit its use. Furthermore, animal-derived matrices are not suitable for translational applications due to risks of immunogen and pathogen transfer. Here, we report a novel zwitterionic polycarboxybetaine (PCB) 3D culture matrix for CRPC-NEPC organoid culture. Zwitterionic materials are known to be superhydrophilic and utilized in various biomedical applications for anti-fouling purposes. As synthetic materials, many properties of culture matrices such as mechanical stiffness, degradation rate, type, and concentration of cell-adhesion peptides, can be fine-tuned based on different organoid disease models. The prostate cancer organoids cultured in an optimized PCB hydrogel group incorporated with fibronectin-mimicking peptides with higher stiffness express similar levels of marker mRNAs and proteins, while showing a higher proliferation profile and larger organoid diameter, indicating that CRPC-NEPC organoids exhibit a more pronounced preference for our pure zwitterionic culture system. Additionally, qPCR results also show that the expression levels of marker genes in the PCB group are less differentially expressed than that in the Matrigel group after multiple passages, indicating that the zwitterionic culture system is giving less background noise to organoids during culture due to its superhydrophilicity and lack of many unwanted peptides. This zwitterionic culture system can offer great opportunities in various organoid disease models and subsequent applications.

502f Biodegradable Small Polymersomes with Encapsulated Indocyanine Green J-Aggregates for Photoacoustic Cancer Imaging

Keith P. Johnston, Mohammed Kawelah, Ceren Atila Dincer, Sangheon Han, Aasim Hussain, Alexander Marras, Thomas Truskett, Konstantin V. Sokolov

Imaging plays a critical role in all stages of cancer care from early detection to diagnosis, prognosis and therapy monitoring. Recently, photoacoustic imaging (PAI) has started to emerge into the clinical realm due to its high sensitivity and ability to penetrate tissues up to several centimeters deep. Herein, we encapsulated indocyanine green J-aggregates (ICGJ), one of the only FDA-approved exogenous contrast agents that absorbs in the near-infrared range, at high loadings up to 40% w/w within biodegradable polymersomes (ICGJ-Ps) composed of poly(lactide-co-glycolide-b-polyethylene glycol) (PLGA-b-PEG). The small Ps hydrodynamic diameter of 80-90nm will enhance the ability to penetrate tumors, while conjugation with EGFR targeting cetuximab antibodies results in high specificity in breast cancer cells. Even when exposed to serum, the 14 nm thick membrane of the Ps prevents dissociation of the encapsulated ICGJ for at least 48 hours with a high level of near-infrared (NIR) absorbance (I_{890}/I_{785} ratio of approximately 5.0) to generate strong PA signals. The PA signal of polymersome-labeled cancer cells increased 3.5-5-fold at 895 nm excitation as a function of the level of EGFR expression on the breast cancer cells, indicating excellent selectivity. The cells were successfully detected in tissue mimicking phantoms with PAI in proportion to their quantity up to a depth of 5 mm, given the high ICGJ loading using a linear array transducer. This demonstrates excellent sensitivity that is clinically relevant to applications such as a tumor margin detection, evaluation of lymph nodes for presence of micrometastasis, and laparoscopic imaging procedures. In the future, in vivo evaluation of ICGJ-Ps will be conducted to illustrate the utility of these polymersomes for tumor detection and therapy monitoring.

562a The Impact of Nanomedicine: 30,000 Orthopedic Nano Implants with No Failures and Still Counting

Thomas Webster

Nanomedicine is the use of nanomaterials to improve disease prevention, detection, and treatment which has resulted in hundreds of FDA approved medical products. While nanomedicine has been around for several decades, new technological advances are pushing its boundaries. For example, this presentation will present an over 25 year journey of commercializing nano orthopedic implants now in over 30,000 patients to date showing no signs of failure (Figure 1). Current orthopedic implants face a failure rate of 5 – 10% and sometimes as high as 60% for bone cancer patients. Further, Artificial Intelligence (AI) has revolutionized numerous industries to date. However, its use in nanomedicine has remained few and far between. One area that AI has significantly improved nanomedicine is through implantable sensors. This talk will present research in which implantable sensors, using AI, can learn from patient's response to implants and predict future outcomes. Such implantable sensors not only incorporate AI, but also communicate to a handheld device, and can reverse AI predicted adverse events (Figure 1). Examples will be given in which AI implantable sensors have been used in orthopedics to inhibit implant infection and promote prolonged bone growth. In vitro and in vivo experiments will be provided that demonstrate how AI can be used towards our advantage in nanomedicine, especially implantable sensors. Lastly, this talk will summarize recent advances in nanomedicine to both help human health and save the environment.

Figure 1: a) Nanotextured Spinal Implants Now in Over 30,000 Humans with No Cases of Failure. The industry standard is 5-10% implant failure, up to 60% implant failure for cancer patients. b) Nanosensors Now Being Used to Detect the Presence of Bacteria, Inflammatory Cells, or Bone-forming Cells on Implants. Such nanosensors can communicate to handheld devices and can respond in real-time to reverse adverse events. Such sensors have been implanted in sheep for up to 12 weeks showing the ability to promote bone growth.

a)

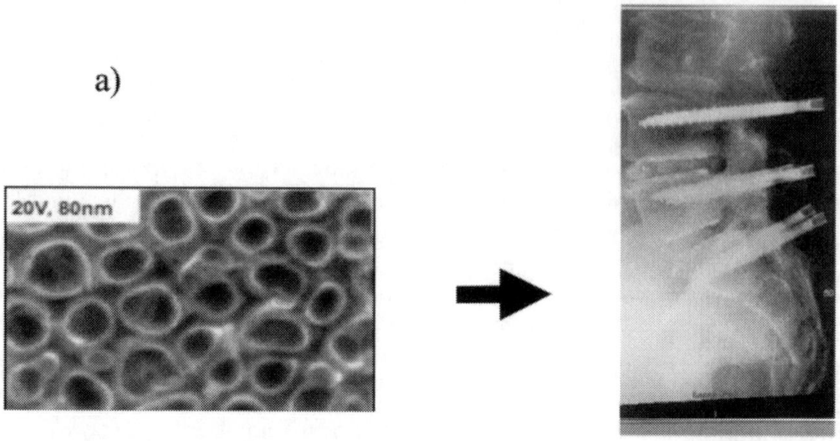

b)

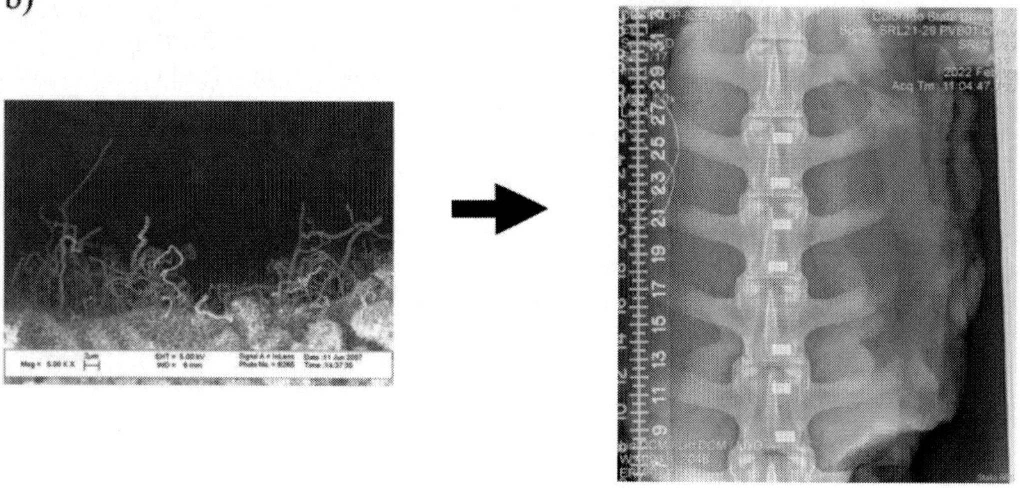

562b Nanothermometry-Guided Robotic and Minimally Invasive Laser Tissue Soldering

Oscar Cipolato, Tobias Leuthold, Marius Zäch, Georg Männel, Sam Aegerter, Calinda Sciascia, Alexander Jessernig, Sima Sarcevic, Jachym Rosendorf, Vaclav Liska, Dennis Kundrat, Romain Quidant, Inge Herrmann

Laser tissue soldering combines laser light with a specialized adhesive material called solder paste, offering a compelling alternative to traditional sutures and staples. This innovative surgical technique primarily utilizes protein-based adhesives to form robust, watertight bonds. Its suitability for minimally invasive procedures presents distinct advantages over conventional methods, particularly in delicate tissues like blood vessels, nerves, and the intestine, where complications such as inflammation, infection, and fluid leaks can arise.

However, the effectiveness of laser tissue soldering heavily relies on precise temperature management during the procedure. Our research introduces a novel approach to address this challenge, employing controlled laser tissue soldering guided by fluorescent nanothermometers. These nanothermometers, seamlessly integrated into protein-based solders, enable non-invasive temperature monitoring. As a significant step toward more accurate and less invasive soldering techniques, we demonstrate automatic temperature control and intelligent laser power modulation for achieving optimal outcomes. This progress is facilitated by incorporating an intelligent solder paste containing near-infrared (NIR) fluorescent nanothermometers and nanoabsorbers such as gold nanorods or titanium nitrate nanoparticles.

Moreover, our study integrates computer vision-based solder recognition and various strategies to guide surgery by analyzing the temperature distribution based on nanothermometry during soldering, thereby enhancing procedural precision. We also emphasize the crucial advantages of minimally invasive soldering in clinically relevant scenarios.

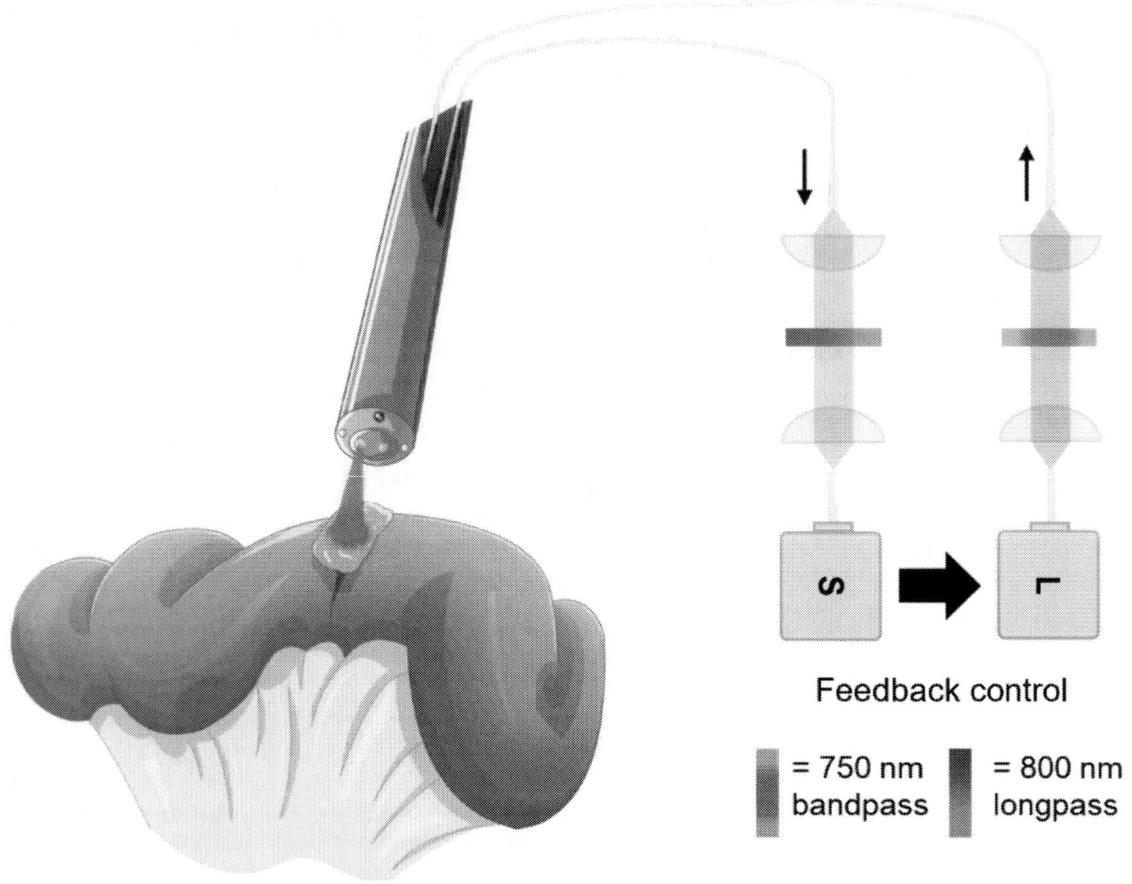
S
L
Feedback control
= 750 nm bandpass
= 800 nm longpass

562c Combinatorial Development of Nebulized Formulations for Pulmonary mRNA Delivery

Allen Jiang, Jacob Witten, Idris Raji, Robert Langer, Daniel G. Anderson

Inhaled delivery of mRNA using lipid nanoparticles (LNPs) has the potential to treat a wide variety of diseases. Nebulized delivery of mRNA to the lungs faces several unique physical and biological challenges. This includes stability during nebulization, where shear forces imparted on the formulation cause LNPs both to release encapsulated mRNA as well as to aggregate. Additionally, the lung epithelial microenvironment poses substantial cellular and extracellular barriers to delivery. The dense mucus matrix covering the lung epithelium represents a significant transport barrier to reaching epithelial cells. Following transport across the mucus layer, epithelial cells in the lungs are extremely difficult to transfect apically owing to evolutionary pressures to resist endocytosing potential pathogens inhaled into the lungs. Here, we systematically addressed both the physical and biological barriers to inhaled mRNA-LNP delivery to yield a highly effective inhaled mRNA delivery approach for the lungs.

To address the challenge of nebulizer instability, we first utilized a design of experiment screening methodology to identify incorporation of a cationic helper lipid which promoted greater electrostatic complexation of mRNA within the LNP to prevent nebulizer-induced mRNA loss. We further enhanced nebulizer stability by rationally designing the LNP nebulization buffer composition. By reducing the pH of the nebulization buffer to protonate ionizable lipids within the LNP formulation, we were able to increase LNP surface charge and improve colloidal stability through electrostatic repulsion. The addition of branched polymeric excipients into the LNP formulation further prevented aggregation during nebulization through steric hindrance. Having developed a generalizable LNP formulation that was stable to nebulization, we next addressed the challenge of intracellular delivery to the lung epithelium. We synthesized and screened a large, novel library of biodegradable ionizable lipids using fully differentiated air-liquid interface cultured primary lung epithelial cells to identify several excellent lung-transfecting hits. Our final combination of novel ionizable lipid, charge-stabilized formulation, and stability-enhancing excipient yielded a state-of-the-art formulation for mRNA delivery to the lung epithelium in mice. Furthermore, repeat dosing of our optimized LNP formulation was well tolerated and did not result in loss of efficacy. By addressing the physical and biological bottlenecks to nebulized mRNA delivery, we developed LNP formulations that are excellent candidates for applications from pulmonary vaccination to protein replacement therapy.

562d Production of Stable, Low Cost mRNA Lipid Nanoparticle (LNP) Powders By Confined Impinging Jet Precipitation and Cold Spray Drying: Stability and Process Design

Robert K. Prud'homme, Mark Kastantin, Kurt Ristroph

Lipid Nanoparticles (LNPs) that deliver mRNA and siRNA have dramatically impacted the field of therapeutic drug delivery. Over 3 billion doses of the COVID vaccine produced by Pfizer based on our confined impinging jet (CIJ) technology have been administered to date. A significant limitation of mRNA LNPs has been the need to distribute and hold these LNPs at -40 to -80 C (cold chain distribution). This has limited the ability to distribute these vaccines in global health applications to less developed nations. We present a new approach to produce stable dry powders by scalable and economic processes. The mRNA LNPs are produced by CIJ precipitation and then dry powders are produced by "cold spray drying". This approach produces dry powders at significantly lower cost than lyophilization and by a process that is scalable. It also eliminates the freezing step, which is a major source of instability in current COVID vaccines that requires high concentrations of lyoprotectants. The fundamentals of each process step will be presented.

562e Kinetics of Capture of Gram-Positive Bacteria By Polydopamine-Coated Magnetic Nanoparticles

Bowen Houser, William Pitt, Alyson N. Camacho, Camille A. Bryner, Tochukwu P. Okonkwo, Rajendra Gautam, Roger G. Harrison, Karine Chesnel

In this study, we explored the rapid capture of pathogenic bacteria using polydopamine (pDA)-coated magnetic nanoclusters (MNCs). While pDA is typically reported as anti-bacterial and anti-biofilm-forming in other publications, we observed strong attractive interactions between pDA-MNCs and several Gram-positive bacteria species—including the pathogenic bacteria *Streptococcus mutans,* usually seen in dental caries, and the opportunistic pathogens *Staphylococcus epidermis* and *Staphylococcus aureus.*

The MNCs were synthesized using a solvothermal process consistent with previous publications, producing clusters of nanoparticles that are about 120 nm in diameter as measured by scanning electron microscopy. A pDA coating was synthesized upon the surface of the MNCs and estimated to be 5 nm thick with transmission electron microscopy.

The kinetics of bacterial capture were measured by mixing pDA-coated MNCs with a bacterial suspension, collecting the MNCs with a strong magnet after various incubation times, and then measuring how many bacteria remained in suspension (not collected by MNCs) after each incubation time. When fit to a mathematical model, the kinetic data for S. mutans suggests there is first-order capture with respect to both bacteria and MNC concentration at short times. Further research is being conducted on the kinetics of *S. epidermidis* and *S. aureus.*

These kinetic details can be used to scale processes for bacterial capture in applications in biomedical diagnostics, and in food and water quality assessment.

562f Radioluminescent Nanoparticles for Radiation-Induced Photodynamic Therapy in Cancer

Sung-Ho Shin, Dhushyanth Viswanath, You-Yeon Won

Photodynamic therapy (PDT) is a promising cancer treatment modality involving photosensitizer activation by light to produce cytotoxic species like reactive oxygen species (ROS) or singlet oxygens ($^1O_2$)[1]. While 5-aminolevulinic acid (ALA) PDT is effective for superficial lesions[2,3], its use is limited in deep-seated tumors due to light penetration constraints[4,5]. To address this, we developed PEG-PLA-coated CaWO4 nanoparticles (PEG-PLA/CWO NPs) as energy transducers for X-ray radiation-induced PDT (RT-PDT)[6]. These NPs generate ROS upon X-ray exposure and emit blue radioluminescence[7], activating ALA-derived protoporphyrin IX (PPIX) to induce $^1O_2$ production, mimicking ALA-PDT effects[8]. In vivo evaluation using PEG-PLA/CWO NPs and ALA prodrugs in a mouse tumor model demonstrated their efficacy in deep-seated tumors. However, nanoparticle distribution in tumors is limited by the extracellular matrix (ECM)[9,10], prompting the development of collagenase-functionalized PEG-PLA-coated CWO nanoparticles (Col-PEG-PLA/CWO NPs). Collagenase facilitates NP penetration through the ECM, significantly improving their spreading within tumors[11]. Micro CT analysis revealed a sevenfold enhancement in NP distribution volume with collagenase functionalization. Evaluation of RT-PDT efficacy using Col-PEG-PLA/CWO NPs and ALA prodrugs in the mouse tumor model demonstrated superior tumor growth suppression compared to unfunctionalized NPs. These findings highlight the potential of collagenase-functionalized NPs to improve PDT outcomes in deep-seated tumors.

References

(1) Lucky, S. S.; Soo, K. C.; Zhang, Y. Nanoparticles in Photodynamic Therapy. *Chemical Reviews* **2015**, *115* (4), 1990-2042, Review. DOI: 10.1021/cr5004198.

(2) Dolmans, D. E.; Fukumura, D.; Jain, R. K. Photodynamic therapy for cancer. *Nat Rev Cancer* **2003**, *3* (5), 380-387. DOI: 10.1038/nrc1071.

(3) de Haas, E. R.; de Vijlder, H. C.; Sterenborg, H. J.; Neumann, H. A.; Robinson, D. J. Fractionated aminolevulinic acid-photodynamic therapy provides additional evidence for the use of PDT for non-melanoma skin cancer. *J Eur Acad Dermatol Venereol* **2008**, *22* (4), 426-430. DOI: 10.1111/j.1468-3083.2007.02445.x.

(4) van Straten, D.; Mashayekhi, V.; de Bruijn, H. S.; Oliveira, S.; Robinson, D. J. Oncologic Photodynamic Therapy: Basic Principles, Current Clinical Status and Future Directions. *Cancers (Basel)* **2017**, *9* (2). DOI: 10.3390/cancers9020019.

(5) Ishizuka, M.; Abe, F.; Sano, Y.; Takahashi, K.; Inoue, K.; Nakajima, M.; Kohda, T.; Komatsu, N.; Ogura, S.; Tanaka, T. Novel development of 5-aminolevurinic acid (ALA) in cancer diagnoses and therapy. *Int Immunopharmacol* **2011**, *11* (3), 358-365. DOI: 10.1016/j.intimp.2010.11.029.

(6) Viswanath, D.; Won, Y. Y. Combining Radiotherapy (RT) and Photodynamic Therapy (PDT): Clinical Studies on Conventional RT-PDT Approaches and Novel Nanoparticle-Based RT-PDT Approaches under Preclinical Evaluation. *ACS Biomater Sci Eng* **2022**, *8* (9), 3644-3658. DOI:

10.1021/acsbiomaterials.2c00287.

(7) Jia, S. C.; Ge, S. F.; Fan, X. Q.; Leong, K. W.; Ruan, J. Promoting reactive oxygen species generation: a key strategy in nanosensitizer-mediated radiotherapy. *Nanomedicine* **2021**, *16* (9), 759-778, Review; Early Access. DOI: 10.2217/nnm-2020-0448.

(8) Kriska, T.; Korytowski, W.; Girotti, A. W. Role of mitochondrial cardiolipin peroxidation in apoptotic photokilling of 5-aminolevulinate-treated tumor cells. *Arch Biochem Biophys* **2005**, *433* (2), 435-446. DOI: 10.1016/j.abb.2004.09.025.

(9) Holback, H.; Yeo, Y. Intratumoral drug delivery with nanoparticulate carriers. *Pharm Res* **2011**, *28* (8), 1819-1830. DOI: 10.1007/s11095-010-0360-y.

(10) Dolor, A.; Szoka, F. C. Digesting a Path Forward: The Utility of Collagenase Tumor Treatment for Improved Drug Delivery. *Mol Pharm* **2018**, *15* (6), 2069-2083. DOI: 10.1021/acs.molpharmaceut.8b00319.

(11) Warwick, D.; Arandes-Renú, J. M.; Pajardi, G.; Witthaut, J.; Hurst, L. C. Collagenase Clostridium histolyticum: emerging practice patterns and treatment advances. *J Plast Surg Hand Surg* **2016**, *50* (5), 251-261. DOI: 10.3109/2000656X.2016.1159568.

562g Nicotinamide-Loaded Nanopeptoids for Energy Regeneration to Drive DNA Repair in Neonatal Brain Injury

Hui Du, Hoang Trinh, Renyu Zheng, Haoyu Wang, Kylie Corry, Olivia Brandon, Tommy Wood, Chunlong Chen, Elizabeth Nance

Background: Nicotinamide adenine dinucleotide (NAD) is a coenzyme found in all living cells that plays a crucial role in various cellular processes, including cell metabolism, energy production, and DNA repair. Neonatal brain injury results in DNA damage, which activates poly(ADP-ribose) polymerase-1 (PARP-1) and decreases cytosolic nicotinamide adenine dinucleotide (NAD+), decreasing adenosine triphosphate (ATP) production, impairing mitochondrial function, and inactivating cellular metabolism, leading to cell death. Increasing NAD+ levels could therefore restore cellular NAD+ to aid in DNA repair and increase cell survival. In the brain, NAD+ is more likely to be made by nicotinamide (NAM)-derived nicotinamide mononucleotide (NMN) available locally to brain cells. However, mammalian cells cannot import NAD+, and the cell-specific targeting of NAD+ and NAM is limited. Therefore, cellular delivery strategies are needed to capture the potential of an NAD+-restorative therapeutic approach.

Methods: In this study, we developed a nanopeptoid delivery strategy to replenish cellular redox state and energy production following acutely injured, energy-depleted brain cells (Fig. 1A). Peptoids are sequence-specific heteropolymers that were developed as protein mimetics possessing advantages of both synthetic polymers and biopolymers. We self-assembled peptoids into tubular form to create peptoid nanotubes (PNTs) with different functional groups. Varying lengths of PNT were achieved by sonication (Fig. 1B). NAD+ or NAM were electrostatically associated or conjugated to the PNT before sonication, and the drug conjugation stability post sonication was confirmed by measuring released drug concentration via liquid chromatography. Free NAD+ or NAM, blank PNT, and NAM-PNT, or NAD+-PNT at a dose of 20 μg/mL were applied topically to organotypic whole hemisphere (OWH) slices isolated from the postnatal day 10 (P10) rat brain. After 24 h of treatment, healthy and oxygen-glucose deprived (OGD) treated slices, an ex vivo model of neonatal hypoxic-ischemic (HI) injury, were evaluated for cell death and PNT-cellular localization using confocal microscopy. Intracellular ATP, cell death, and cell proliferation were also quantitively analyzed. In a P10 rat model of neonatal HI, NAM-PNT was systemically administered immediately after HI injury at a single dose of 50 μg/mL of NAM (500 mg/kg). Controls included HI pups treated with saline, free NAM, and blank PNTs. Animals were sacrificed 72 h after treatment to evaluate gross injury, area loss, and neuropathology.

Results: Peptoids are usually around 2 μm in length without sonication and around 10 nm in diameter and show tubular structure on atomic force microscopy (AFM). The length can be further reduced to hundreds of nanometers after 2 h of sonication without changing other characteristics. PNTs conjugated with NAD+ or NAM achieved high drug loading efficiency and no cytotoxicity in brain cells (Fig. 1C). NAM-PNTs localize in microglia and associate with neurons (Fig. 1D). NAM-PNTs improve cell viability in response to OGD in OWH brain slices (Fig. 1E). NAM-PNTs replenish intracellular ATP levels by 24h after treatment (Fig 1F) and drive glial proliferation (Fig 1G). We see an associated shift in reduced pro-inflammatory cytokine production and increased anti-inflammatory cytokines. In the P10 HI rat, a single dose of NAM-PNTs administered systemically decreased brain tissue area loss and improved neuropathology.

Conclusions: Our study demonstrates that NAM delivery via PNTs has strong therapeutic potential for cell-specific delivery and energy regeneration in the acutely injured neonatal brain.

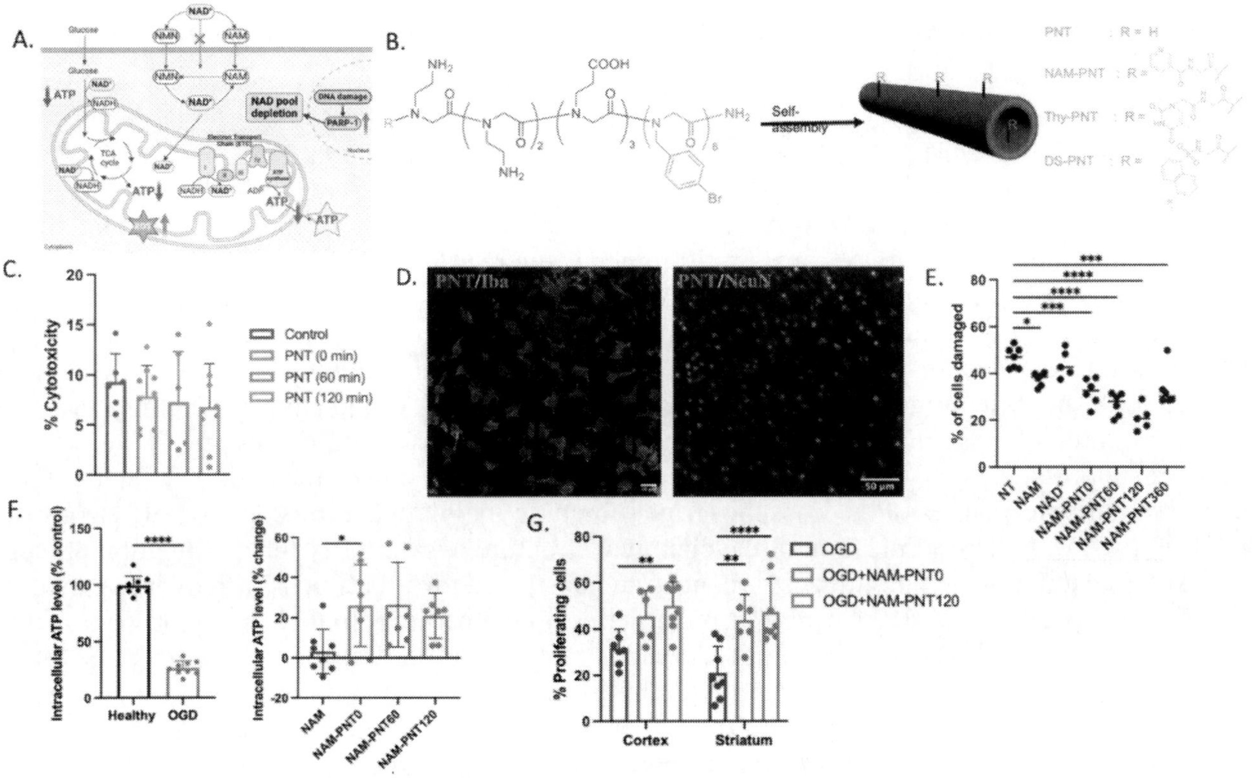

Figure 1. Peptoid nanotube (PNT) delivery of NAM for neuroprotection in the neonatal injured brain. A. Schematic diagram showing the need to deliver NAD+ or its precursors to the injured cells. B. Chemical structure of NAM=PNT and NAD+-PNT. C. Cytotoxicity of blank PNT of varying lengths on healthy slices. LDH release (%) values are normalized to the LDH release of acute slices immediately treated with TX-100. D. NAM-PNTs localize in Iba+ cells, and show association but no clear uptake in NeuN+ cells in an organotypic whole hemisphere. E. Cell viability of oxygen-glucose deprived (OGD) slices in response to free drug treatment (NAM, NAD+), or NAM-PNT treatment with varying PNT lengths, compared to a non-treated (NT) OGD control. Each data is a single brain slice. F. OGD decreases intracellular ATO and NAM-PNTs are able to increase intracellular ATP following treatment after OGD. G. NAM-PNTs drive cellular proliferation in the OGD-exposed brain. (* denotes P<0.05, ** denotes P<0.01, *** denotes P<0.001, **** denotes P<0.0001)

562h Modified AAV Vectors with Phosphoserine-Containing Zwitterionic Peptides for Enhanced Cell-Specific Transduction and Immunosuppressive Capacities

Ruoxin Li, Di Liu, Wenchao Gu, Sijin Luozhong, Meng Cui, Zhefan Yuan, Shaoyi Jiang

Although adeno-associated viruses (AAVs) are considered low immunogenic and safe for gene delivery, the immunogenicity of capsids still impedes the re-administration of AAVs. The current method using poly(ethylene glycol)(PEG) to shield AAV from being recognized by neutralizing antibodies efficiently reduces humoral immunity but is not effective against cell-mediated immune responses. Here, we modify the AAV surface with phosphoserine-containing zwitterionic peptides to induce AAV-specific immune tolerance to facilitate multiple AAV injections. The phosphoserine moiety is known for specific binding to the receptors on monocytes and induction of immunosuppressive environment. Moreover, our zwitterionic design of the peptide balances the excessive anionic charges of the phosphoserine moiety to maintain the transduction efficiency of the modified AAV. After optimization of the chemical structure, we simplify the conjugation process into a one-step reaction which minimizes virus loss and sustained virus activity. Moreover, we evaluate the binding specificity of the phosphoserine moieties on the peptides to their receptor via a surface plasma resonance (SPR) assay. We further demonstrate the similar transduction and uptake efficiency of modified AAV compared to wildtype AAV in different cell lines and in vivo at a single injection to the animal. Finally, in the setting of multiple injections, we reveal that the modified AAV has better transduction efficacy and induced lower antibody titers than the wild type. Overall, this immunosuppressive zwitterionic peptide presents a new platform for engineering AAV vectors to enable re-administrations and increased delivery efficiency.

562i Engineered Virus-like Vesicles for In Vivo Targeted Drug Delivery in TNBC Treatment

Yi-Chih Tsai, Nadine Elkasri, Kai Su-Greene, Simian Cai, Ann Li, Vaibhav Upadhayay, Aixin Shi, Hayleigh Goodrich, Dorin Artzi, Sijin Luozhong, Qiuming Yu, Richard Cerione, Wenchao Gu, Shaoyi Jiang

Triple-negative breast cancer (TNBC) is among the most challenging types of cancers to treat, due to the absence of receptor expression, including estrogen receptor, progesterone receptor, and human epidermal growth factor receptor 2, which are common targets for hormone therapies. Chemotherapy remains the primary treatment for TNBC; yet, its efficacy is limited by drug resistance, severe side effects, and high toxicity levels. Exploring alternative therapeutic approaches, scientists are targeting the 'glutamine addiction' and metabolic shift known as the Warburg Effect, where TNBC cells rely heavily on glutamine for energy. By inhibiting glutaminase (GLS), the enzyme essential for converting glutamine into glutamate, this strategy aims to starve TNBC cells of a critical resource for their growth. Despite the promise of GLS inhibitors, their poor water solubility leads to suboptimal potency, bioavailability, and circulation time, compromising their effectiveness in suppressing tumor growth. While various carriers have been developed to overcome drug delivery limitations, challenges such as high cytotoxicity, rapid clearance, and poor targeting capacity persist.

Addressing this challenge, our study utilized leukocyte-derived extracellular vesicles (EVs) known for their natural ability to target tumors. We aim to directly deliver hydrophobic GLS inhibitors to the tumor's inflammatory environment. Despite the biocompatibility and innate delivery capability of leukocyte EVs, effective encapsulation of drugs and vesicle uptake by recipient cells remain significant hurdles. Here we genetically modified donor cells to include virus-like capsid proteins inside the EVs to encapsulate drugs, along with envelope proteins on the EV surface to enhance cellular uptake by target cells. Additionally, we introduced an RNA stabilizer to form hydrophobic pockets within these capsids, significantly enhancing the ability of the EVs to enclose and maintain GLS inhibitors. These inhibitors were simply added to the culture of donor leukocytes, which then naturally incorporated them into the capsids within the EVs they produced. To evaluate the efficacy of engineered EVs in carrying GLS inhibitors, we used fluorescent inhibitors and vesicle lipid membrane dyes for easy tracking and visualization within recipient cells. More importantly, to assess its effectiveness against TNBC in vivo, we established a murine xenograft model by injecting human TNBC cells into immune deficient mice, closely replicating the human disease environment. We then administered the engineered EVs intravenously in a controlled dosage schedule to ensure maximum impact with minimal side effects, monitoring tumor growth and assessing the treatment's effectiveness through caliper measurements and final analysis of tumor sizes and weights after euthanizing the mice.

As a result, we demonstrated that engineered vesicles enriched within the tumor, successfully encapsulated, and transported hydrophobic GLS inhibitors into TNBC cells. Our in vivo studies demonstrated that systemic delivery of engineered EVs significantly reduced the size of TNBC tumors. Specifically, mice treated with EVs loaded with GLS inhibitors showed a 53% decrease in tumor volume and a notable extension in lifespan, highlighting the effectiveness of this novel delivery system in addressing TNBC. This new approach leverages the unique virus-like features of engineered nanocarriers, which mimic natural viral mechanisms of cellular entry and infection, to enhance therapeutic delivery and efficacy significantly. By integrating virus-like components into EVs, our research paves a new pathway for the encapsulation and targeted delivery of hydrophobic drugs into the inflammatory microenvironments of various pathological conditions, such as cancers and neurodegenerative diseases.

AUTHOR INDEX

AIChE
120 Wall Street, FL 23
New York, NY 10005-4020

ISBN 979-8-3313-1687-7

CURRAN ASSOCIATES INC.
proceedings.com